BIOETHICS

Bioethics

A Coursebook

Compost Collective

https://www.openbookpublishers.com

©2025 Daan Kenis, Mayli Mertens, Franlu Vulliermet, Varsha Aravind Paleri, Yanni Ratajczyk, Emma Moormann, Christina Stadlbauer, Bartaku Vandeput, Nele Buyst, Lisanne Meinen, Kristien Hens, Ina Devos and Ilya Gordon Villafuerte (Compost Collective)

This work is licensed under a Creative Commons Attribution-NonCommercial 4.0 International (CC BY-NC 4.0). This license allows you to share, copy, distribute and transmit the text; to adapt the text for non-commercial purposes of the text providing attribution is made to the authors (but not in any way that suggests that they endorse you or your use of the work). Attribution should include the following information:

Attribution should include the following information:

Compost Collective, *Bioethics: A Coursebook*. Cambridge, UK: Open Book Publishers, 2025, https://doi.org/10.11647/OBP.0449

Further details about CC BY-NC licenses are available at http://creativecommons.org/licenses/by-nc/4.0/

All external links were active at the time of publication unless otherwise stated and have been archived via the Internet Archive Wayback Machine at https://archive.org/web

Digital material and resources associated with this volume are available at https://doi.org/10.11647/OBP.0449#resources

Information about any revised edition of this work will be provided at https://doi.org/10.11647/OBP.0449

ISBN Paperback: 978-1-80511-510-6
ISBN Hardback: 978-1-80511-511-3
ISBN PDF: 978-1-80511-512-0
ISBN HTML: 978-1-80511-514-4
ISBN EPUB: 978-1-80511-513-7

DOI: 10.11647/OBP.0449

Cover image by Bartaku Vandeput.
Cover design by Jeevanjot Kaur Nagpal

Contents

AUTHORS BIOGRAPHIES ... ix

1. BIOETHICS: A GLOBAL APPROACH
Introduction ... 1
Our approach to bioethics: an ethico-onto-epistemology ... 3
Philosophical method ... 8
Conclusion ... 12
Bibliography ... 12

2. MORAL THEORIES
Introduction ... 15
Utilitarianism ... 16
Deontological ethics ... 18
Care ethics ... 22
Conclusion ... 25
Bibliography ... 25

3. ENVIRONMENTAL ETHICS
Introduction ... 27
Deep ecology ... 32
Ecofeminism ... 32
Environmental justice ... 34
Conclusion ... 40
Bibliography ... 41

4. HEALTH CARE ETHICS
Introduction ... 45
Doing (medical) ethics ... 46
Medical and clinical ethics: the patient-physician relationship ... 50

Ethics of medical AI	53
Reproductive ethics	58
Public health ethics	61
Research ethics in biomedical research	64
Conclusion	70
Bibliography	70

5. ANIMAL ETHICS AND ANIMAL EXPERIMENTATION

Animal ethics	73
Animal experimentation	77
Conclusion	86
Bibliography	86

6. EPIGENETICS

Introductory remarks	89
Introduction to epigenetics	89
Ethics of epigenetics	94
One finding, many ethical and political claims	99
Parental responsibility in epigenetics	101
Conclusion	106
Bibliography	106

7. SYNTHETIC BIOLOGY

What is synthetic biology?	109
Conceptual issues in SynBio	110
Ethics of SynBio	112
SynBio and non-dualism	120
Conclusion	123
Bibliography	124

8. LITERARY BIOETHICS

Introduction	127
Literary form and genres	128
Genre and different media	129
(Bio)ethical questions in fiction	131
Conclusion	135
Bibliography	136

9. BIOETHICS AND (BIO)ART

What is BioArt?	139
The ethics of BioArt	140
BioArt and bioethics	141
How BioArt can contribute to ethical considerations towards the 'unknown'	142
Conclusion	144
Bibliography	144

INDEX   145

Authors Biographies

Nele Buyst is a doctoral researcher in Philosophy. She is currently working on an interdisciplinary project that uses the metaphor and practice of kintsugi to think about the concept of repair and to explore the potential of aesthetic practices to heal modern relations to surroundings. She is interested in feminist arts-based research methods, ecofeminism, and posthumanist philosophy. Alongside her research, she publishes poetry and is an editor of *Rekto:Verso,* a magazine for culture and critique.

Ina Devos is a doctoral researcher in Philosophy with a background in biotechnology and bioethics. Their current research focuses on the ethics of human proteomics, building on the fields of proteomics, bioethics, data ethics, and research ethics. They also have a specific commitment towards interdisciplinary research, collaborating with scholars of proteomics, bioinformatics, data science, philosophy of science, and science and technology.

Ilya Gordon Villafuerte is a research master's student currently working on his doctoral application. He is mostly engaged with social ecology and relational ethics in urban contexts, and is a student assistant for this course.

Daan Kenis is a doctoral researcher in Philosophy with a background in philosophy and pharmaceutical sciences. His research tackles questions at the intersection of philosophy of science and bioethics in the context of data-intensive healthcare, precision medicine, and molecular biology. His work employs insights from feminist philosophy of science, philosophy of science in practice, and social epistemology to address normative concerns in biomedical practice and healthcare.

Kristien Hens is a full professor in Bioethics at the University of Antwerp. Her research interests include biological concepts, ethics of psychiatry, ethics of postgenomics, and environmental ethics. She has a particular interest in microbes and what they can teach us. She wrote *Chance Encounters. A Bioethics for a Damaged Planet* and *Towards an Ethics of Autism.*

Lisanne Meinen is a postdoctoral researcher with a background in cultural studies and philosophy. Her doctoral research focused on the cultural representation and phenomenological understanding of neurodivergence in and through videogames. She is currently developing a new research project on immersive technology and disability.

Mayli Mertens is a Marie Skłodowska-Curie Fellow at the Philosophy department of the University of Antwerp and the Founding Director of the Atlas Bioethics Center in San Jose, Spain. She investigates how sense-making, through human and artificial cognition, impacts the physical world. Her main scientific interest is in epistemology and global bioethics. She teaches on bias, critical thinking, and technological innovation.

Emma Moormann is a postdoctoral researcher in Philosophy. Her PhD thesis discusses moral responsibility distributions in the context of epigenetics. She has since worked on a variety of projects related to (bio)ethics, including the philosophy of (step)parenthood, the concept of resilience in neurodiversity, and disability theory. Starting from 2025, she is project manager of the interdisciplinary project *Death Care*.

Varsha Aravind Paleri is a doctoral researcher in Philosophy. With a strong background in biotechnology and bioethics and extensive experience as a molecular biologist, her current doctoral research focuses on the ethics of synthetic biology. Her scholarly interests lie at the intersection of technological and environmental ethics, with a particular focus on understanding and applying non-Western philosophical frameworks—mainly Indian Hindu philosophy—to address complex ethical challenges in these domains.

Christina Stadlbauer is an artist-researcher with a PhD in Chemistry and graduate studies from Apass, a post-master program for advanced performance and artistic research. Her practice explores relationships between humans and other-than-human life forms—including plants, animals, bacteria, and mycelium—seeking to re-negotiate our environmental connections. Working at the intersection of arts, sciences, and philosophy, Stadlbauer presents her research through installations, performances, rituals, and curated events.

Joke Struyf is a doctoral researcher working on the frictions between normative motherhood, mothering, feminism, and the sense of maternal agency. Before this, she analysed young Europeans' ideas about gender, fertility treatments, and parenthood; and together with neurodivergent participants, she explored the challenges of co-creation in a big European project on neurodiversity. Her main influences are Black feminism and queer, crip, and decolonial theory and activism.

Yanni Ratajczyk is a postdoctoral researcher in Philosophy at the University of Rijeka and the University of Antwerp. His research primarily focuses on moral philosophy, the philosophy of Iris Murdoch, and the intersection between ethics and aesthetics. He has published on narrative identity, moral creativity (the topic of his PhD), moral imagination, moral perception, and the philosophy of Iris Murdoch.

Bart H. M. Vandeput (BE/FI), known as Bartaku, is an artist and researcher specializing in Bioart and intermedia art. His transdisciplinary practice explores the human condition through collaborations with plants, microbes, light, energy, and technology, integrating diverse entities, media, and disciplines into installations, interventions, exhibitions, talks, and lectures. He currently directs a research project on microbiomes in the cooling towers of nuclear facilities, involving artists and researchers from four universities in Belgium and Finland.

Franlu Vulliermet is a doctoral researcher in Philosophy working on a normative account of environmental relationships in the context of pollution and epigenetics. His research engages with a variety of schools of thought, including non-Western perspectives informed by his time spent living with Indigenous populations in the Ecuadorian Rainforest. Prior to this, he was a research associate at INSEAD and Harvard Business School.

1. Bioethics: A Global Approach

Introduction

This is a coursebook about bioethics. Bioethics is the branch of applied ethics that studies the philosophical, social, and legal issues arising in medicine and the life sciences. It is a discipline that is nowadays often conflated with biomedical ethics, as opposed to environmental ethics. In this book, however, we consider bioethics in the spirit of Van Rensselaer Potter, who thought bioethics should be a global endeavour (Potter, 1988). By that, he meant that it should truly be an ethics of life, spanning both the human and the other-than-human world: both biomedical ethics and environmental ethics. Also, for us—the writers of this textbook—it does not make much sense to consider duties towards human health and the environment separately. Recent developments in molecular biology, such as epigenetics and exposomics, demonstrate that such separation is unhelpful. Recent initiatives such as *OneHealth* link human, other-than-human, and ecosystem health and advocate for a transdisciplinary approach to health. Such a transdisciplinary approach not only involves biological and biomedical scientists but also includes sociological, legal, ethical, and political perspectives. This book aims to give a broad overview of the method of bioethics and some of the central debates in the field. It is aimed at students of philosophy, biology, biomedical sciences, bioengineering, and all those interested in researching and working with life in the widest sense.

As a student in the sciences, finding yourself with this coursebook, you might wonder, 'Why would scientists have to learn about ethics?' Is something scientifically accurate not immediately equivalent to what is morally good, and is doing good science in the technical sense not the same as doing morally good science? Acting ethically correct is 'doing good', and do we not all know intuitively what that is? Maybe conceptions about what is good are different for everybody or based on the culture in which we are situated. Maybe finding a universal answer to the question 'what is good' makes no sense. In this book, we will ask ourselves questions such as: what is good ethical practice in general and for the life sciences in particular? Which ethical dilemmas do life scientists face? It will become clear from the start that we do not aim to provide straightforward answers to ethical dilemmas. Instead, we want to offer you the tools to reflect on ethical issues and arrive at a balanced conclusion.

We shall start with a small exercise: Imagine that you are a biologist working on chemical herbicides to inhibit undesirable plants in agriculture, forestry, or non-crop areas such as industrial sites, roadsides, and lawns. What kind of ethical questions do you think may arise?

Arthur Galston was a biologist and plant physiologist at Yale University, who turned into a bioethicist later in his career. His dissertation on the flowering process of soybean plants led others to develop Agent Orange: the most widely employed herbicide during the Vietnam War, used to defoliate forests and eliminate enemy cover and food sources. Galston, as a bioethicist, said, "The only recourse for a scientist concerned about the social consequences of his work is to remain involved with it to the end" (Galston, 1972). While well-intended, we would like to offer two points of reflection. Firstly, we can expect a scientist's work's social and ethical consequences to continue long after the scientist has passed away. Secondly, proper critical thinking requires reflection before and during scientific development, too, not just after. In the following paragraphs, we discuss three aspects of basic critical reflection.

Know what you don't know

Our knowledge production and our methods to test hypotheses are, by definition, limited by our senses and the instruments we have to extend what we can observe with those senses. The human naked eye cannot capture ultraviolet or infrared light, but technology allows us to measure and characterize these frequencies. Our ears do not capture radio signals without the help of a radio. There are plenty of phenomena like these that we can only observe through technology. However, there are many more phenomena which we do not have the tools required to capture. Sometimes, we can infer from the feedback we receive. For instance, there are many toxins we cannot smell or taste. But if we consistently drop dead after inhaling or ingesting them, we can assume that they are deadly. Perhaps we can also find other ways to detect them. However, we will likely not detect phenomena outside of our field of perception, whether through senses, technology, or feedback. As a result, we will not even know whether these phenomena exist (and yet this does not mean they don't exist!). These boundaries of our empirical knowledge represent the first way in which our view of reality may be limited.

Perspectives and bias

Aside from reducing what humans *can* know, the limitations in our sense-making also strongly *distort* all that we *do* know. Not just *what* we can experience but *how* we can experience it is informed by our senses, technology, and feedback, but also our intuition and cognition. Mental processes have their own limitations, and they are greatly affected by cognitive, cultural, and contextual factors. Heuristics and cognitive biases give shape to the information by influencing the interpretation of that information.

Changing one's point of view often seems to change everything. Even if the situation hasn't changed at all, a different perspective may change not only the most relevant factors, but also the ways one can or wishes to respond to the situation. On the one hand, this explains why we can observe a global trend of questioning and reforming old concepts. On the other hand, it creates extreme, often polarized, viewpoints where people question even the most fundamental assumptions (e.g. fake news, flat earthers).

Outdated and incomplete knowledge

Finally, even if we do have some concrete, 'objective' information, the world changes fast. Scientific knowledge travels slowly, although this, too, is changing. And yet, even though the spread of information is accelerating, schoolbooks can be outdated by the time they come into print. This is especially the case for basic, general knowledge about geography and recent history, for instance (e.g. world population statistics). Moreover, knowledge can still be correct in and of itself, but when new insights are gained on a wider scale, these change the meaning of the smaller parts (e.g. the tip of the iceberg).

Critical reflection requires our awareness of all these limitations, as well as the openness and curiosity to find satisfying answers in spite of non-ideal conditions.

Our approach to bioethics: an ethico-onto-epistemology

When scientists from fields of study such as biology, biochemistry, or biomedicine hear the word 'ethics,' they often think about procedural requirements—such as GDPR, ethics committees, and consent forms—or restrictions on the use of laboratory animals, and safety regulations. Indeed, part of the role of bioethicists is to sit on committees that evaluate the ethics of research proposals and (clinical) interventions. Specifically, these evaluations take the form of assessing whether consent forms are clear enough about the risks that research participation entails, or whether animals are unnecessarily harmed in experiments. But it is also the task of ethicists, in their capacity as philosophers, to think about what it *means* that animals are not unnecessarily harmed. How do we define what 'harm' entails? And when is harm 'acceptable'? How do we weigh the interests of other-than-human animals vs humans? And this is where it gets interesting.

Our approach to bioethics is reminiscent of what Oxford philosopher Onora O'Neill once said. O'Neill (born 1941), who received an honorary doctorate from the University of Antwerp on 3 April 2021, has called bioethics "a meeting ground for a number of disciplines, discourses and organisations concerning ethical, legal and social questions raised by advances in medicine, science and biotechnology." (O'Neill, 2002, p. 1). Indeed, one cannot think about the practical dilemmas in the life sciences without engaging with these sciences. Bioethicists are often philosophers, but also sometimes people with a background in the life sciences. Galston, encountered earlier, is but one example of a biologist turned bioethicist. However, this 'interdisciplinarity' is not only applicable to dialogues between ethicists and scientists. Ethics and bioethics necessarily also engage with other branches of philosophy.

Traditionally, philosophy has been divided into practical and theoretical disciplines, each with subdisciplines with their own journals and conferences. Subdisciplines of theoretical philosophy include metaphysics, philosophical anthropology, and epistemology. In the former two disciplines, people ask themselves what human beings are, what the world is, and what the universe is. In epistemology, people ask what knowledge is and how we know things. Practical philosophy includes disciplines such as ethics, political philosophy, and social philosophy. In political philosophy, people consider questions of politics and power: structures and ideologies such as capitalism, democracy, colonization, communism, and the patriarchy are analyzed. In social philosophy, people consider the origins and essence of a society and the relation between the individual and social structures.

At the same time, we may question this carving up of philosophy (and science in general) into different subdisciplines. As we will see, bioethicists need to coordinate with philosophers of science, metaethicists, and scientists. Questions about how the world is (ontology) and how we know things (epistemology) are intimately entangled with ethical questions. The way bioethics is approached in this coursebook is hence inspired by the concept of *ethico-onto-epistemology*, first coined by feminist philosopher of science Karen Barad (Barad, 2007). In this neologism, you may recognize several of the main subdisciplines in philosophy: ontology, epistemology, and ethics. Let us discuss these in order.

Ontology

Ontology refers to our conceptualization of reality and of the phenomena that are part of reality. Specific concepts we take for granted in everyday life are not straightforward upon closer consideration. Think, for example, about 'curing a disease'. What do we actually mean when we call something a *disease*? Do we mean that there is a specific biological *cause*, such as in the case of influenza? Do we mean that a particular person significantly deviates from the statistical mean, such as with high blood pressure? Does this relate to the way people typically function?

This type of question is also relevant for researchers. For example, much research is spent searching for genes that explain the cause of autism. At the same time, we can ask ourselves why people are doing this kind of research. Do we eventually want to 'cure', 'solve', or 'prevent' autism? That idea itself is problematic, as the neurodiversity movement has argued. Should we consider autism a disease or a disorder, or is it just a variant of typical human behaviour? Thinking about how concepts are used in scientific disciplines is also essential for communicating science. If scientists find a statistical association between a specific gene and a specific behaviour or characteristic, such as intelligence, can we say 'the gene for...' has been found? What's more, different cultures may have different views on reality and may even be said to 'live in different worlds'. For example, Indigenous peoples in North and South America view humans as deeply interconnected with the land, animals, and spirits, whereas other cultures view the land and its creatures as resources to be used. This already shows how views on the world are always connected with what we consider morally good.

Epistemology and philosophy of science

Epistemology is the branch of philosophy that is concerned with how we know things. The philosophy of science, which is considered by some a subbranch of epistemology, critically reflects on many questions in science and tries to clarify these. Think, for example, about the question of *scientific knowledge*. When is knowledge scientific? Is a statement scientific if we have sufficient empirical proof? If we are thinking about the question 'what is reality made of?', for example, we often stumble upon concepts that we may never be able to prove empirically. Think about string theory in physics. We may use strings to explain certain observable phenomena in reality, but we will probably never be able to observe the strings themselves empirically. Is string theory science, or is this the place where the distinction between philosophy and exact science blurs? Philosophers of science and physicists can think together about scientific practice and what counts as scientific fact in light of these new theories.

Another common assumption pertains to science and scientific knowledge as progressing linearly and cumulatively. Philosophers of science have investigated this idea and have tried to consider whether and how scientific progress is possible. Thomas Kuhn, a philosopher of science, challenged this widely held view. Kuhn argued that scientific development is not a smooth, continuous process but rather occurs through revolutions or paradigm shifts (Kuhn, 2012). According to Kuhn, scientists operate within shared frameworks or *paradigms*, and a crisis arises when anomalies—results that do not fit the current framework—challenge the prevailing paradigm. This leads to a scientific revolution where a new paradigm emerges, fundamentally changing the way scientists understand, approach their field, and even see the world. Different paradigms, Kuhn suggests, are *incommensurable*. Since scientists in different frameworks understand the world differently, we can establish no shared measures to compare different paradigms (e.g. the Ptolemaic vs Copernican worldview). In this view, it becomes difficult to see scientific progress as purely cumulative.

In addition to the question of the possibility of scientific progress, feminist philosophers of science have also explored how scientists are also (subconsciously) led by other influences. They have suggested that social and political influences play a decisive role in research, separately from the principles and norms of scientific practice itself (i.e. the 'scientific method'). Societal, political, and economic interests, for one, inform what questions scientists ask and the types of research we pursue. We have historically been, for example, more interested in male health than in women's health. The hype surrounding the Human Genome Project also led to a 'geneticization' of research: framing research questions in primarily genetic terms.

While these factors primarily inform science from the 'outside', values also inform scientific decision-making in the lab. A famous example of the 'internal' influence of values on science is the *argument from inductive risk*, which suggests that in conditions of uncertainty—endemic to all science—the acceptance or rejection of a particular hypothesis always involves the risk of getting it wrong. Whether we accept or reject a hypothesis thus not only depends on the data but also on our weighing of the *consequences of error*. A factory making lightbulbs can, for example, tolerate a 5% error rate; whereas, in cancer diagnostics, we might want to be more conservative. Therefore, while scientists generally agree upon a specific threshold for statistical significance (often $p<0.05$)—i.e. for the acceptance or rejection of a hypothesis—the judgement should depend on how we value the consequences of getting it wrong.

If values play a significant role in deciding what we find interesting and how scientists come to scientific knowledge, this raises the question of whose values we take into account: who is at the helm of scientific decision-making? Who gets to decide what types of research we pursue and how we pursue them? Feminist philosophers of science have noted that there exists no neutral, value-free standpoint from which we (and also scientists) approach phenomena. Instead, we are all deeply *situated*. This means that what we can know, which evidence we have access to, and how we weigh those data are all (variously) dependent on our social identity. Arguing against the common-sense view, feminist epistemologists suggest that identity does matter in science, and diversity in the scientific community can itself lead to better knowledge acquisition. They suggest that those who are often excluded from the research community or those affected by the topic of interest often hold valuable and novel insights into the issue at hand. The efforts of feminist activists in tandem with the inclusion of women in biomedical research, for example, led to significant improvements in women's health, a topic which had been considered relatively unimportant. The inclusion of diverse standpoints, these authors suggest, is not only morally favourable but can also lead to better, more objective science. In sum, what is considered scientific depends on current scientific paradigms and what we find acceptable as a society, and who is contained within this 'we'.

Ethics

What is ethics?

Ethics is a branch of philosophy that deals with morality on different levels. There are two non-normative branches. In *descriptive ethics* or *moral sciences*, morality is approached from social sciences, psychology, and cultural anthropology. For example, when discussing care ethics, we will talk about psychologists Lawrence Kohlberg (1927–1987) and Carol Gilligan (born 1936), who have studied the different stages of moral development in children. *Metaethics* is a branch of ethics that investigates why human beings are moral and how they are moral. Philosophers look at history, the social sciences, or biology to understand why human beings have moral sensitivity. For example, we can ask ourselves if it is sufficient to feel guilty to be moral or whether you must have a rational conviction that you have done something wrong. Is morality a matter of emotions or reason? Metaethics also studies concepts such as good and evil and justice.

Ethics also has two normative branches. In *general normative ethics*, philosophers think about which kind of behaviour is good or bad. Ethicists try to lay down the basic principles of morality in rational terms and look for an encompassing moral theory. In applied ethics, these questions are asked in specific contexts. Specific moral dilemmas from specific subdomains of human action are analyzed and specified. For example, in business ethics, we can determine the extent of a company's responsibility concerning the well-being of its employees and their families. In media ethics, we investigate journalists' duties towards those interviewed. Bioethics, the topic of this coursebook, is also a form of applied ethics.

Metaethics: why are we sensitive to morality?

As a scientist, you are maybe specifically interested in one question from metaethics: why are people sensitive to morality, and is this specific to human beings? One explanation stresses the struggle to survive and states that ethics primarily applies to humans in a community. Seventeenth-century philosopher Thomas Hobbes (1588–1679) situates the origins of morality in *egoistic prudence* (1996). He asks us to imagine that at a certain moment there were only a few people and lots of food and other resources, at the beginning of human history. However, as the population grew, people had to compete for those resources. Individuals were entangled in a bitter struggle to survive. Only the strongest made it. In this harsh climate, the social contract emerged: people realized that it was to everyone's advantage to keep to a set of moral rules and norms. These rules and norms were institutionalized in laws and enforced by the state.

More recently, Western scientists have acknowledged that other-than-human animals also show altruistic behaviour. Biologists Sarah Brosnan and Frans de Waal demonstrated that sensitivity to fairness and altruism, considered prerequisites for morality, is also present in other-than-human animals (2003). Of course, many people consider human morality to be different from morality in other-than-human animals. There is a place for religion in human morality, for taboo, which might not be the case with other-than-human animals. Still, de Waal and others have argued that morality has its roots in our animal nature, contrary to what Hobbes has suggested.

Metaethics: how do moral facts and scientific facts relate to one another?

When trying to solve ethical dilemmas about new technologies, people often feel that it is sufficient to list these technologies' benefits and disadvantages or risks. What is morally good can, so to say, be discovered by looking at scientific data and be logically deduced from the facts. The underlying idea is that moral facts can be reduced to non-moral facts. Moral facts thus have no separate ontological status in reality. This is called *ethical* (or *moral*) *naturalism*. An example of an ethical naturalist is Peter Railton (born 1950). According to Railton, an act is morally good only if the act is done by a fully rational and informed subject (a subject that has 'looked at the data') but also takes the social point of view into account and includes all interests of all involved. Hence, it is necessary to look at empirical data to understand the concept of moral goodness (Railton, 1986). *Ethical* (or *moral*) *non-naturalists* believe that the good cannot be reduced to other, non-moral facts. The most important name associated with ethical non-naturalism is G. E. Moore (1873–1958). He states that if moral goodness coincides with a natural characteristic—for example, what is good is pleasurable—then whether a specific act that increases pleasure is good becomes a senseless question because the answer would be, per definition, positive—just as the question of whether bachelors are unmarried would make little sense. For Moore, however, the question about the goodness of acts does make sense. It is essential to ask the question. Hence, goodness is a fundamental, separate characteristic that cannot be directly deduced from natural facts. It follows that the properties of goodness cannot be defined but can only be shown and grasped. Goodness is that which our moral intuitions point to, not what we can imply from empirical data (Moore, 1903).

For this coursebook, we shall not go into further detail concerning discussions between ethical naturalists and non-naturalists, nor will we take a stance about who is right. When confronted with bioethical questions, we must consider scientific facts as part of ethical deliberation. For example, people may intuitively feel that we should not genetically modify plants (the 'yuck feeling'), but—especially in a field like bioethics—it is vital to look at and thoroughly understand the scientific facts and advantages that such technologies may yield. However, this does not mean we can deduce good and bad from a mere risk-benefit analysis. Evaluative judgements unavoidably creep into such risk-benefit analysis itself (recall the argument from inductive risk), and judging whether or not a specific use of technology is morally good requires something more.

Philosophical method

People often think that philosophy is a reflection of one's personal values. However, philosophy is a discipline in which one tries to think clearly and thoroughly about certain things. In this way, philosophy differs from other forms of critical thought in several ways:

The method of cases (thought experiments)

Rather than generating scientific data through experiments, philosophers develop philosophical theories based on data from their areas of interest. They often use *thought experiments*. Thought experiments are fictional cases with which one tries to test or bring to the fore certain philosophical intuitions. Some of these thought experiments are even funny. For example, philosopher Derek Parfit (1942–2017), who wrote extensively about the concept of personal identity, uses the example of a 'teletransporter' in his book *Reasons and Persons (1984)* (Parfit, 1984). This teletransporter is similar to the one in *Star Trek*. If you get into the teletransporter, you fall asleep, get destroyed, and break down into atoms. This information is then copied to Mars, where you are reassembled. Is this newly reassembled person the same as you were on Earth? What if the original person has not been destroyed, but copies of you are made throughout the entire universe? What does it mean to be and to remain the same person?

Such thought experiments may seem far-fetched, but they can help us solve dilemmas closer to home. They help us reflect on what it means to have an identity, and how stable such an identity is. For example, the question of how much you have to be altered neurologically before you become a different person is relevant to several bioethical issues. Is it ethical to perform euthanasia on a person with dementia if this person has, at a time when they did not yet have dementia, expressed the wish to be euthanized if they ever develop dementia? What about drugs like *Ritalin*, which can make life easier but may also influence one's personality? Is keeping your 'personal identity' an essential value, or is our identity held together by the story we tell ourselves and others? The use of thought experiments is not only reserved for philosophers, by the way. Modern physics has also often started from thought experiments, such as the double-split experiment by Thomas Young (1804).

During the last twenty years, there has been some critique within philosophy on using thought experiments 'from the armchair'. Thought experiments often draw certain conclusions or state philosophical (and universal) truths. However, is this possible from 'the armchair'? How do we know that our intuitions are the correct intuitions? Are they not relative to the culture we have grown up in? For example, think about the thought experiments inspired by Edmund Gettier (1927–2021), the so-called Gettier cases (1963). They deal with the circumstances under which you can know something. In epistemology, it is often stated that you really know something if (1) it is true, (2) you believe yourself that it is true, and (3) you are justified in your belief that it is true. This is the *justified true belief* thesis of knowledge and was long considered the standard account. Edmund Gettier used a thought experiment to prove that this conception of knowledge is actually not complete: something more than justified true belief is necessary to constitute knowledge.

> ### Gettier's thought experiment ('Smith and Jones')
>
> Smith and Jones have both applied for a job. Smith has good reasons to think that Jones (1) will get the job and (2) has ten coins in his pocket. Jones has shown him the coins, and Smith has counted them. Moreover, the assistant of the HR director came out after the job interviews and told Smith that Jones would get the job. Hence, Smith has good reasons to believe the following statement: Jones will get the job, and Jones has ten coins in his pocket. From this, it follows that he is also convinced of the following statement: the person who will get the job has ten coins in his pocket. But the company's people have changed their mind at the last moment and will offer the job to Smith.
>
> Moreover, Smith also has ten coins in his pocket, although he is unaware. Hence, the first statement that Jones will get the job and has ten coins in his pocket is not true. But this is the statement from which Smith has deduced that the person who will get the job has ten coins in his pocket. Intuitively, we feel that Smith does not really *know* that the person who gets the job has ten coins in his pocket, although it is a true and justified belief. Hence, the intuitions called forth by this thought experiment suggest that it is not enough to have a truly justified belief in order to be able to speak about really *knowing* something.

Experimental philosophy

However, is it really true that everyone feels intuitively that we cannot talk about really knowing something in the case of Smith? Or are philosophers from their armchairs the only ones who think like this? Since the beginning of this century, some philosophers have started to deploy methods from psychology and sociology to test longstanding philosophical intuitions. This *experimental philosophy* often questions the function of thought experiments and the philosophical intuitions they are thought to invoke. Experimental philosophers demonstrate, for example, that philosophical intuitions can differ between cultures. A study by Weinberg, Nichols, and Stich (2001) has suggested that people from East Asia have a different intuition regarding such Gettier cases compared to Americans: in some scenarios, they would consider that Smith 'really knows' this. Other studies found no difference. Some people have criticized experimental philosophy. They say it is not real philosophy but rather psychology. The research would use a flawed methodology (poor sample size, data analysis, etc.). Nevertheless, it is also interesting for philosophers to relate to empirical data, either by doing the research themselves or by being informed by empirical studies. Moreover, experimental philosophy demonstrates that the values and thoughts that have formed the gist of Western philosophy may be less universal than previously thought. In this course, we mainly talk about Western philosophers. This does not mean Western philosophers have a more direct line to the truth: we must remain aware that philosophy and ethics are also partially culturally sensitive endeavours—or, as we put it earlier, philosophical and ethical reasoning are *situated*. Later in this chapter, we will introduce concepts such as *Buen Vivir* and *decolonizing ethics*.

Empirical and embedded bioethics

A similar concern pertains to theories and frameworks operationalized by bioethics. As we will see in the chapter on health care ethics, a ubiquitous tool in bioethical reflection are the principles of autonomy, beneficence, non-maleficence, and justice. It is often assumed that these principles are indicative of a 'common morality' and shared across culture and time. Given what we have seen in the section on feminist philosophy of science, it is important to also critically approach those intuitions. One way to achieve that is through empirical methods.

Empirical bioethics is a field that uses empirical methods, such as surveys, interviews, or observations, to inform normative questions about relevant bioethical topics. Bioethicists use empirical methodologies to question and explore the implications of bioethical decisions, to describe and assess attitudes towards specific intuitions, or to assess normative assumptions with regard to specific technologies or medical practices. A typical empirical bioethics paper will consist of survey or interview results of a relevant stakeholder group—physicians, researchers, patients, citizens—on a particular matter of concern. A common (and familiar) concern with the field is that the samples in empirical bioethics research are often quite homogeneous. In line with what we read on the feminist philosophy of science, we should note that if the preferences and intuitions of stakeholders are meant to inspire additional moral considerations, it is important to sample a sufficiently diverse group of individuals.

Embedded ethics is another somewhat novel methodological approach to applied ethics. Embedded ethics responds to concerns over ethics' increasing distance from scientific practice. This distance is reflected in scientists' concerns that bioethics hampers scientific progress, is often irrelevant to the actual science, and tends to focus on sensationalized cases such as designer babies and human cloning. Ethicists, too, increasingly recognize the limitations of a top-down approach to ethics that decontextualizes and abstracts away the specificities of the actual cases and responsibilities that (individual) scientists may have in their work. This leads—among other concerns—to issues of translating ethical principles to the practice and personal experiences of scientists. In turn, ethicists have been increasingly calling for an *embedded* approach to bioethics. Embedded ethics refers to the ongoing practice of integrating ethics in the entire (scientific) process. Several levels of embedding have been suggested in the literature, ranging from better ethics education in the sciences, the implementation of ethics throughout scientific research (from research design and planning to implementation), or the integration of an ethicist in the lab. Taking inspiration from fieldwork in field philosophy and anthropology, and the idea that good science is ethical science and vice versa, embedded ethics aims to bridge the gap between science and ethics (and scientists and ethicists) by inspiring dialogue, collaboration, and deliberation on ethical issues as they arise within scientific practice. Other ethicists are critical of the so-called 'embedded turn', arguing that it reduces ethical inquiry to a mere servant of scientific progress. Embedded ethics may stand too close to science and not have sufficient distance to engage critically with disruptive technologies and scientific developments.

Conclusion

In this chapter, we have drawn on Van Rensselaer Potter's vision to present bioethics as a unified and transdisciplinary approach. Through the example of Arthur Galston's involvement in the development of Agent Orange, we demonstrated how ethical reflection must be embedded in scientific practice from the start. We adopt an ethico-onto-epistemological approach, emphasizing that ethics, knowledge, and our understanding of reality are deeply intertwined. Drawing on philosophy of science (e.g. Kuhn's paradigms) and feminist epistemology, this approach challenges the idea of science as value-free and highlights the importance of diversity in knowledge production. This chapter has also explored major branches of ethics, including metaethics and applied ethics, contrasting ethical naturalism with non-naturalism. We have introduced thought experiments as a method of philosophical reasoning, while also discussing experimental philosophy's efforts to test intuitions empirically. Finally, we have presented empirical and embedded bioethics as practical approaches for integrating ethics into real scientific contexts, showing how bioethics can—and should—be grounded in both philosophical reasoning and everyday research practice.

Bibliography

Barad, Karen Michelle. 2007. *Meeting the Universe Halfway: Quantum Physics and the Entanglement of Matter and Meaning*. Durham: Duke University Press.

Brosnan, S. F., and F. B. M. de Waal. 2003. "Monkeys Reject Unequal Pay". *Nature* 425 (6955): 297–299. https://doi.org/10.1038/nature01963

Galston, A. W. 1972. "Science and Social Responsibility: A Case History". *Annals of the New York Academy of Sciences* 196 (4): 223–35. https://doi.org/10.1111/j.1749-6632.1972.tb21231.x.

Gettier, Edmund. 1963. "Is Justified True Belief Knowledge?". *Analysis* 23 (6): 121–23. https://doi.org/10.1093/analys/23.6.121

Harding, Sandra. 1991. *Whose Science? Whose Knowledge?: Thinking from Women's Lives*. Ithaca: Cornell University Press. https://www.jstor.org/stable/10.7591/j.ctt1hhfnmg

Hobbes, Thomas. *Leviathan*. 1996. Edited by Richard Tuck. Cambridge: Cambridge University Press.

Kuhn, Thomas S. 2012. *The Structure of Scientific Revolutions: 50th Anniversary Edition*. Edited by Ian Hacking. Chicago: University of Chicago Press. https://press.uchicago.edu/ucp/books/book/chicago/S/bo13179781.html

McLennan, Stuart, Amelia Fiske, Daniel Tigard, Ruth Müller, Sami Haddadin, and Alena Buyx. 2022. "Embedded Ethics: A Proposal for Integrating Ethics into the Development of Medical AI". *BMC Medical Ethics* 23 (1): 6. https://doi.org/10.1186/s12910-022-00746-3

Moore, G. E. 1903. *Principia Ethica*. Cambridge: Cambridge University Press.

"One Health Commission". n.d. Accessed 10 July 2024. https://www.onehealthcommission.org/

O'Neill, Onora. 2002. *Autonomy and Trust in Bioethics*. Cambridge: Cambridge University Press. https://doi.org/10.1017/CBO9780511606250

Parfit, Derek. 1987. *Reasons and Persons*. Vol. I. Oxford: Clarendon Press.

Potter, Van Rensselaer. 1988. *Global Bioethics: Building on the Leopold Legacy*. East Lansing: Michigan State University Press.

Railton, P. (1986). "Moral Realism". *The Philosophical Review* 95 (2): 163–207. https://doi.org/10.2307/2185589

Railton, Peter. 2003. *Facts, Values, and Norms: Essays toward a Morality of Consequence*. Cambridge Studies in Philosophy. Cambridge: Cambridge University Press. https://doi.org/10.1017/CBO9780511613982

Weinberg, Jonathan M., Shaun Nichols, and Stephen Stich. 2001. "Normativity and Epistemic Intuitions". *Philosophical Topics* 29 (1–2): 429–60. https://doi.org/10.5840/philtopics2001291/217

Young, Thomas. 1804. "The Bakerian Lecture: Experiments and Calculations Relative to Physical Optics". *Philosophical Transactions of the Royal Society of London* 94: 1–16.

2. Moral Theories

Introduction

Moral theories are general theories that try to answer the questions 'what is morally good?' and 'how can we distinguish good from bad?' In applied ethics, we often use principles that we can trace back to these theories. Therefore, it is vital that we know moral theories and what the weak points of these moral theories are. In the following chapter, we discuss four approaches: utilitarianism, deontology, virtue ethics, and care ethics. But before we start, a warning is in order. When confronted with certain ethical questions from a specific field of study, such as biomedicine, we usually do not pick out our favourite theory and apply it. Although some ethicists consider themselves as strictly 'utilitarian' or 'deontologian'—this view is called *ethical absolutism or monism*— ethical questions are often too complex and too diverse to be solved with one theory. For example, ethical questions can pertain to policymaking, interpersonal relations, or both. Moreover, we can ask ourselves whether there is always one good answer to an ethical dilemma. This does not mean that there cannot be better and worse answers in any case. At the same time, knowing the different aspects of morality through these moral theories, and the challenges that each approach has, helps you look at a specific ethical question through different angles.

Courses on applied ethics, such as bioethics, often start with an overview of moral theories and principles, as does this book. This ubiquity of moral theories might give the impression that they represent opinions about which underlying framework of morality is the best. Indeed, that may be the way they were conceived in the first place. It is more helpful, however, to view them as describing *aspects of morality*. As such, they are valuable tools: you should try to look at ethical dilemmas using different approaches and theories and know the drawbacks to each approach. Moreover, it is essential to realize that a specific type of reasoning may be utilitarian or deontological, but that does not mean that applying utilitarianism or deontology to a specific ethical dilemma yields straightforward answers. We will see some examples later in this chapter.

Utilitarianism

Imagine that you are standing beside tracks, looking at a trolley approaching. You notice that there are five people tied on the tracks. If the trolley continues its course, it will kill these five people. However, you can ensure that the trolley will get onto another track by pulling a particular lever. There is one person tied to this alternative track. This person will be killed if you pull the lever. What should you do? This thought experiment was initially described in 1967 by British philosopher Philippa Foot (1920–2010). It seems reasonable to answer here that it is better to cause the death of one person than the death of five persons. If you want to find out what the best course of action is, looking at the consequences looks like a good place to start. However, it may be that pulling the lever does not sit well with some of us. Aren't we actively killing that one person, and is killing not a bad thing no matter what? What if that one person is very young and the five other ones are very old? Or if that one is a loved one, such as a friend or child?

A moral theory that evaluates actions based on the consequences of these actions is *consequentialism*. The best-known version of consequentialism is *utilitarianism*. Utilitarianism states that an act is good if it results in 'the greatest good for the greatest number of people'. Hence, it is a question of cost-benefit analysis: you weigh up what it would cost to do something and the consequences. This idea, although very old, was first systematized in 1789 by Jeremy Bentham (1748–1832). In the spirit of Enlightenment, Bentham states that faith is secondary to reason: moral rules should not come from God, but you should deduce them by thinking properly. He proposed a *hedonistic calculus* to find out what the good consequences are. Only pleasure matters: what is good is what causes us pleasure, and what is immoral is what causes us pain. Pleasure (or pain) can be measured in intensity, length, certainty, and whether they are followed by opposite emotions.

However, many philosophers question the use of pleasure as a basis for cost-benefit analysis. Robert Nozick (1938–2002) was one of them. In *Anarchy, State and Utopia* (2013), he describes a thought experiment called *the experience machine:* imagine that there is a machine that gives you pleasurable experiences. According to the hedonistic approach, the mere *experience* of pleasure should be equivalent to the *proper acts* that give you pleasure. Intuitively, however, Nozick states that people would prefer to actually do the acts that give pleasure rather than merely experience the pleasure in itself. Hence, hedonism is based on a wrong assumption.

John Stuart Mill (1806–1873) describes the utility principle as follows: "Actions are right in proportion as they tend to promote happiness, wrong as they tend to produce the reverse of happiness, i.e. pleasure or absence of pain" (Mill, 2001, p. 7). He considered some actions to be qualitatively better than others. For example, pleasures of the mind (intellectual activity) are more important than physical activity (sport). In the twentieth century, people tried to solve problems with the hedonistic interpretation of the utility principle by stating that good has to be defined based on the satisfaction of preferences rather than the provision of pleasure (*preference utilitarianism*). Preferences are not always purely hedonistic—some prefer, for instance, to spend time caring for people in need over going out and enjoying life.

Another further elaboration of utilitarianism was factoring rules into the calculus. *Act utilitarianism* demands that each act should be treated separately. *Rule utilitarianism* demands that an act be judged based on more general rules to maximize happiness. It may be, for example, acceptable to lie to your partner when you have cheated on them because the truth would hurt them and cause them pain. However, rule utilitarianism suggests that breaking the rule 'do not lie' would undermine the maximization of happiness in the long run. It may ultimately lead to a breach of confidence. Lying should not be tolerated because it undermines the foundations of society and leads to less well-being.

Utilitarianism has good points. When it started gaining popularity, it was ethically progressive for its time because it counts everyone who can experience happiness as relevant for moral consideration, including women and children. Jeremy Bentham referred explicitly to animals when he stated: "The question is not, 'Can they reason?' nor, 'Can they talk?' but 'Can they suffer?'"(Bentham, 1789, p. 311). Peter Singer (born 1946), a contemporary utilitarian philosopher, uses similar arguments to defend non-human animals. According to him, to make a moral distinction between humans and non-human animals is *speciesism*. Speciesism is analogous to racism and sexism. It means that people attribute different values to different creatures on the basis of their similarity to their own species—in this case, to the human species, which is thus deemed superior.

Utilitarianism appeals to our common sense and is a sound basis for policymaking. When policymakers are confronted with the fact that they have little money available for research, should this money not be spent on diseases that cause the most suffering? Should it not be spent on cancer research rather than Botox treatments for cosmetic purposes? Here, we encounter the weak points of utilitarianism. How can we measure suffering, and who is then suffering the most? How much weight should we give to the suffering of animals? What about foetuses or the biosphere?

Because utilitarianism is about deciding what is right or wrong based on *future* results, it remains, to some extent, *speculative*. We could defend nuclear energy because it produces far less CO_2 than fossil fuels and could, hence, be part of the solution to climate change. The risk of a nuclear disaster might annihilate this benefit. Utilitarianist thinking can also cause suffering and can, specifically, be detrimental for minorities. Indeed, when an action benefits 51% of the population—meaning that their happiness increases—at the cost of a decrease in the happiness of 49% of the population, is this a morally just act? Let us say you are a doctor with five patients, each needing a different organ to survive. You find a lonely and homeless man who is the perfect match to save all five lives. Should you kill this man?

Another often-quoted example is slavery: if enslaving one person benefits many people, can we condone it? The theme of sacrificing some to keep the wider social order is often described in art and stories, such as the short story The *Lottery* by Shirley Jackson, and *The Hunger Games* by Suzanne Collins.

Utilitarianism can also lead to conclusions that some intuitively would consider *supererogatory*. *Supererogatory* means that something is beyond doubt good, but at the same time we feel it asks too much of us. For instance, Peter Singer states the following about poverty: "If it is in our power to prevent something bad from happening, without thereby sacrificing anything of comparable moral importance, we ought, morally, to do it" (1972). This seems straightforward, but he illustrates it with a striking example. Let us suppose you walk next to a lake where a child is drowning. You are morally obliged to save the child. However, let us suppose you walk by a shop where they sell shoes and see a pair of expensive shoes you want to buy, which cost 100$. According to Singer, buying this pair of shoes is equivalent to letting the child drown. In fact, with that money, you could have saved a poor child's life, for example, by investing in mosquito nets. For many, saving on shoes to donate the money would seem supererogatory.

Deontological ethics

The rules in rule utilitarianism (you may not steal, you may not murder, you may not lie, etc.) may look the same as the duties described in this paragraph when discussing *deontological ethics*. The most crucial difference is the *reason* why these rules should be followed. The rules in rule utilitarianism have to be followed because they will eventually benefit society. As we shall see in the following paragraphs, deontology asks us to follow the rules because it is our duty to follow them. In utilitarianism, the intention of the person who follows the rules is not essential. They may do good deeds unconsciously or because they enjoy telling the truth or abstaining from murder. However, for the deontologist, the rules must be followed out of a sense of duty—the *intention* matters.

Deontological ethics is primarily associated with Immanuel Kant (1724–1804). Deontology is about the rights and duties we have as individuals for others. Kant wanted to lay down a way of ethical reasoning based on rational procedures that would apply regardless of human beings' desires or social relations. These moral principles would be the same for everyone. Briefly, his theory goes as follows: Kant assumes that each person has inherent dignity. Each person is rational and free to make their own law autonomously. This dignity and autonomy must be respected at all times. Hence, human beings should be treated as ends, not merely means.

According to Kant, morality is to act according to the *categorical imperative*. This means acting according to generalizable maxims or rules of conduct. Such rules bind every rational creature and are grounded in reason. For example, it may be that we would like to lie sometimes, but we cannot wish that 'you are allowed to lie' would become a universal law. These ethical laws are valid for everyone. They are universal and not specific to a particular context, in contrast with *conditional hypothetical imperatives* ('if you want to be cool, you have to buy fashionable clothes'). (Kant, 1997)

Deontological ethics has good points. Ideas such as *respect* and *autonomy* are not obviously present in utilitarian thinking, although many people think they should take a central spot in the theory of morality. Utilitarianism values good consequences; good intentions are less important. Still, the fact that you do something because you feel it is your duty, regardless of the consequences, seems intuitively to be part of morality.

However, the idea of universality in Kantian ethics has been criticized by Charles Mills (1951–2021), who points out that, for Kant, the concept of a rational person did not generalize to women and people of colour. Mills contends that this issue is not just a historical anomaly: abstracting from race (Mills is a proponent of critical race theory) excludes from morality the genuine consequences that racial oppression has had—and still has—on people of colour. The idea of 'colourblind' ethics perpetuates this oppression by abstracting away from the different starting positions based on race. Mills nevertheless still inscribes himself in a Kantian tradition—as the title of his paper, '*Black Radical Kantianism*', indicates. For him, respect for the other should consider how they, as part of a minority group, were—and are—disadvantaged. Failing to do that creates abstract equality that will perpetuate inequality by entrenching disrespect for minorities and their practices (Mills, 2018).

Strictly adhering to the rules of deontology might also lead to counterintuitive conclusions. Think about the following scenario: your friend is being chased by a murderer with an axe. They are looking for protection at your house. The murderer knocks and asks you whether your friend is in the house. A strict Kantian would have to say that you cannot lie and must respond 'yes'. It is sometimes also difficult to imagine how a categorical imperative can be used in policymaking. Kant would, for example, oppose euthanasia in all circumstances. Still, euthanasia is allowed in several countries (including Belgium) under certain conditions.

Moreover, we may also feel intuitively that emotions, not a mere sense of duty, are essential to morality. Who you would consider to be the most moral person in the following scenario: person A, who truly loves taking care of people and spends their time as a volunteer in a centre for the homeless; or person B, who would rather spend their free time watching Netflix but still volunteers in a centre for the homeless because they feel morally obliged? A strict Kantian might vote for person B. Still, we feel that emotions are essential. Maybe person A is the most moral because they have an exemplary character. Can duty in itself be a ground for moral action? Do we need something more, like an inclination to *care for others*?

The American philosopher Christine Korsgaard (born 1952) has added a dimension of *sensemaking* and *personal values* to deontological ethics (Korsgaard, 1996). She offers us a modern interpretation of Kant's thoughts. Therefore, she distinguishes between the 'universal law' and the 'moral law', whereby the categorical imperative is only a formulation of the universal law. It is not the source of 'doing one's duty'. The moral law prescribes content related to your identity, to your being an embodied creature in everyday practice. This does not mean that you are merely guided by desires. In the process of reflective enforcement, you choose to identify yourself with a particular principle or law. This presentation of yourself is your practical identity, a description that makes executing what you want to do worthwhile. You take up a role. What you think you have to do is a conception of what you are. Morality is about 'being able to live with yourself'. In Korsgaard's view, deontological and virtue ethics, which we will discuss later, are coming closer together.

We have seen that the central notion of a rational being in Kant's ethics could become problematic. For Kant, being a rational person grounds who has rights and who has not. Later deontological approaches have tried to extend that to other entities. Foetuses and people with an intellectual disability, for instance, come to mind. And what about other-than-human animals? The contemporary philosopher Tom Regan (1938–2017) has applied a deontological approach to animal rights (Regan, 2004). He stated that animals also have interests, are 'subjects of a life', and have intrinsic worth. Hence, they are 'somebody' rather than 'something'. Still, this approach can be considered speciesist, to some extent. We can imagine that dolphins, whales, and primates are people as well. But what about, for example, invertebrate animals? Are jellyfish part of our moral communities?

Taylor Swift: hardcore deontologist?

"Taylor Swift has made several statements and taken actions that suggest deontological ethical thinking, particularly around artists' rights and ownership of their work. Her stance that artists should own their masters and her decision to re-record her albums reflects a categorical view that certain principles (like artistic ownership) should be upheld regardless of consequences. She has framed these positions in terms of absolute rights and duties rather than purely utilitarian outcomes."

This is a statement generated by the LLM Claude AI in January 2025, in response to our request for an example of a deontologist in pop culture.

In 2021, following a dispute with her record label, Taylor Swift started releasing her old albums as "Taylor's Versions." More than remasters, these versions were a statement of the author's independence in a musical industry where labels own most of the artists' copyrights. In accordance with deontological trains of thought, Claude stresses that such principles should be upheld regardless of consequences. But is that really possible? Taylor Swift, one of the richest and most powerful artists, had the opportunity to claim this independence and rerecord

> her albums herself—a freedom most artists do not have. Swift also received higher royalties for her music as a result.
> Reflect on this case and discuss:
> - What do you think of Claude's stating that Taylor Swift is a deontologist?
> - From a deontological perspective, do (or should) rich and powerful people have higher duties?
> - What would Kant have done (had he abided by his own principles) in Swift's shoes?

Virtue ethics

Utilitarianism and deontology try to lay down formulas for what is good and evil (respectively, 'the greatest good for the greatest number' and 'what you can want as a universal law'). As such, they do not fill out specific prescriptions on how to act in concrete circumstances. *Virtue ethics* tries to do this. Virtue ethics was initially conceived by the Ancient Greek philosopher Aristotle (384–322 BC) in his *Ethica Nicomachea*. Also, in the Middle Ages, virtue ethics was an important line of thinking. For instance, Thomas van Aquino (1225–1274) reconciled Aristotle's virtue ethics with Christian doctrine. He laid down seven virtues: belief, hope, love, wisdom, courage, temperance, and justice. In the twentieth century, virtue ethics experienced a revival with philosophers like Elizabeth Anscombe (1919–2001) and Alasdair MacIntyre (born 1929).

Being virtuous is having a specific sensitivity to what is right under specific circumstances. It is choosing the suitable (or golden) mean between two extremes. For example, a virtuous person is courageous—the middle ground between brutal, unthoughtful behaviour and cowardice. A virtuous person is modest, which is the middle ground between shyness and impudence. These virtues are habits rather than duties. They bring *eudaimonia* (happiness, the good life) to those who have them. You are virtuous by *phronesis* or practical wisdom. You acquire a virtuous character by being sensitive to a specific situation, by the judgement of character, by living in a society in which one can learn. Virtue ethics then emphasizes a moral character, which can be built through practice and education.

Virtue theory has specific good points compared to utilitarianism and deontology. To begin with, these other theories have been accused of being too legalistic, which means that too much weight is given to following abstract rules or principles while morality is also about building good character. Relatedly, utilitarianism and deontology do not have much to say about the content of moral principles. They claim that what is moral should be what benefits the most people (utilitarianism) or what reasonable people should want to be a universal law (deontology). Hence, they could be considered examples of thin moral theory. Virtue theory is a 'thicker' kind of morality, as concepts such as courage, modesty, etc. are not overlooked (Väyrynen, 2025).

Nevertheless, virtue ethics also has some shortcomings. Aristotle believed that leading a virtuous life would ultimately lead to a happy life (*eudaimonia*). For Aristotle, happiness and the good life were inseparably connected: the good is connected to life's ultimate goal (*telos*). In the twenty-first century, this looks naïve. Moreover, virtuousness depends on proper education and training. Can those raised in dire circumstances who have missed the proper education and guidance still lead a virtuous life? Also, what is seen as virtuous can differ from culture to culture. Depending on where one lives, euthanasia can be seen as an act of mercy or as murder. How can virtue ethics deal with such different attitudes? Virtue ethics looks especially suitable for interpersonal relations, but how do you apply specific moral rules to specific circumstances? Virtue ethics tells you how to live, not necessarily what to do when confronted with a specific ethical dilemma.

The concept of virtue and similar concepts have been present in different cultures throughout history. For example, in the text *Life on the Slippery Earth*, Sebastian Purcell describes a similar concept in Aztec culture (Purcell, 2018). Unlike the individual-focused Western approach, Aztec ethics emphasize managing vices with the help of others and leading a socially rooted life. They saw life as fundamentally slippery, and whether you are happy is equally dependent on luck as it is on your own character. Therefore, the community is important to help you get up when you slip. Aztecs believed that virtue is fostered through community support and continuous moral education rather than individual character alone. They viewed the good life as one that is worthwhile and balanced rather than necessarily happy.

Care ethics

Care ethics is an approach to morality that deals primarily with interpersonal relations. It started as a feminist critique of ethics dominated by abstract principles. Carol Gilligan (born 1936) wrote a critique on the different stages of moral development in children laid down by moral psychologist Lawrence Kohlberg (1927–1987). These six stages begin with morality as mere avoidance of punishment—in accordance with social norms and obedience to the law—through to the sixth stage, when moral reasoning is based on abstract reasoning consistent with universal moral reasoning, reminiscent of the Kantian subject.

Before going into the theory, let us discuss a vignette that Kohlberg used in his experiments. The following dilemma was presented to children to test their reactions:

> A woman was near death from a special kind of cancer. There was one drug that the doctors thought might save her. It was a form of radium that a druggist in the same town had recently discovered. The drug was expensive to make, but the druggist was charging ten times what the drug cost him to produce. He paid $200 for the radium and charged

$2,000 for a small dose of the drug. The sick woman's husband, Heinz, went to everyone he knew to borrow the money, but he could only get about $1,000, which is half of what it cost. He told the druggist that his wife was dying and asked him to sell it cheaper or let him pay later. But the druggist said: "No, I discovered the drug and I'm going to make money from it." So Heinz got desperate and broke into the man's store to steal the drug for his wife. Should Heinz have broken into the laboratory to steal the drug for his wife? Why or why not? (Gilligan, 1993, p. 25–32)

When eleven-year-old Jake was presented with this dilemma, he stated that Heinz should steal the drug. He used logic to calculate that the wife's life was more important than the worth of the drugs for the pharmacist. However, when eleven-year-old Amy was presented with the dilemma, she had a different answer. She did not think Heinz should steal the money but should seek other ways to solve the problem, like borrowing the money. If Heinz steals the money, he may have to go to jail, and nobody would be there to take care of the wife. Hence, she did not approach it as a purely logical puzzle but instead as a narrative of relationships. Kohlberg scored her at a lower level of moral development.

However, in her book *In a Different Voice: Psychological Theory and Women's Development* (1993), Gilligan questioned this hierarchy of moral development and stated that there are two kinds of moral voices. The masculine voice is more logical and individualistic, and the emphasis is on protecting the rights of people. The feminine voice emphasizes protecting interpersonal relationships and taking care of people. This is the *care perspective*. Both perspectives are equally good. One is not more moral than the other; they complement each other.

In the last decades, care ethics has become a popular ethical approach. Famous care ethicists include Joan Tronto (born 1952) and Nell Noddings (1929–2022). Critical issues in this approach include respect for the vulnerability of others, the importance of relations, and ethics as responding to a *need*. Some have considered care ethics a kind of virtue ethics (the caring person as a virtuous person, the interpersonal rather than abstract approach). It is an integral part of bioethics and medical ethics, and looking at ethical issues from a relational perspective is indispensable for good ethical decision-making. For instance, if we want to look at the ethics of prenatal screening and pregnancy termination, we should not only consider utilitarian arguments (who will benefit?) or deontological arguments (at what point in development does a foetus deserve respect because it has become an end in itself rather than a means to satisfy a parental drive?). It is also imperative to investigate the individual experiences of pregnant people. Hence, an approach that considers individual experiences becomes more important with care ethics. Care ethics has also been applied to animal ethics. Other-than-human animals are considered vulnerable beings we should care for and protect.

Some care ethicists have also received criticism as they would present caring as an intrinsic characteristic of women, which overlooks social influences on women's behaviour. However, this criticism is no longer valid if both the logical approach to morality and the relational approach are disconnected from gender but are still seen as complementary aspects of human morality. Another criticism is that responding to someone's vulnerability can conflict with respecting someone's autonomy.

Care ethics, with its emphasis on relationality and personal relations, is popular among feminist bioethicists. It has been compared to other bioethical approaches, such as principlist frameworks that prioritize autonomy and other principles. Consequently, whilst some argue that an ethics of care may conflict with an individual's right to autonomy in medical encounters, it is possible to view respect for autonomy as an integral aspect of care. Care ethics recognizes the importance of relationality and considers the right to self-determination from that perspective. For example, when it comes to euthanasia, a caring approach to autonomy entails empathetically engaging with the person making the request, making them feel supported in their decision, and understanding the concerns of those around them. This approach, championed by philosopher Eva Kittay, underscores the relationality inherent in life (Kittay, 2019). Care is a reciprocal process, allowing *both* those receiving and giving care to flourish. Moreover, those receiving care should implicitly or explicitly endorse the care provided for it to be truly caring. An ethics of care extends beyond virtue ethics for healthcare professionals and parents; it also contains a political dimension, as argued by scholars like Joan Tronto. In her book *Caring Democracies*, Tronto defines caring as follows:

> On the most general level, we suggest that caring be viewed as a species activity that includes everything that we do to maintain, continue and repair our 'world' so that we can live in it as well as possible. (Tronto, 2013, p. 19)

This emphasis on caring clashes with the prevailing approach to policymaking and economics. For example, during the COVID pandemic, it became clear how important care-related professions are. At the same time, they are often undervalued and associated with low compensation. Care ethicists, including Tronto, advocate for reevaluating this perspective and recognizing the centrality of real people's lives in politics. An ethic of care relevant to bioethics transcends human interactions and acknowledges the intricate interconnectedness between humans and various entities, encompassing other-than-human animals, microbes, and the environment. Maria Puig de la Bellacasa, in her groundbreaking work *Matters of Care* (2017), expands on Tronto's notion of care as a complex, life-sustaining network, describing it as inherently ethical and political. Puig de la Bellacasa proposes a concept of posthuman care that surpasses interpersonal and human realms, perceiving care as a pervasive condition permeating the fabric and surfaces of the world. Therefore, care reflects a fundamental reality of human existence and our intricate entanglement with the larger world—an understanding also present in Indigenous knowledge and ecofeminism.

Conclusion

This chapter has introduced four moral theories—utilitarianism, deontology, virtue ethics, and care ethics. Rather than prescribing a one-size-fits-all approach, we emphasized the importance of understanding these theories as tools that illuminate different aspects of morality. Utilitarianism focuses on outcomes and maximizing well-being, but it struggles to measure suffering and can be used to justify harming minorities. Deontology, grounded in duty and rational principles, values intention and respect for persons but can lead to rigid or counterintuitive conclusions. Virtue ethics centres on character and moral development through habituation and context-sensitive judgement, offering a richer picture of morality but raising questions about cultural differences and applicability to specific dilemmas. Care ethics, emerging from feminist critiques, prioritizes relationality, vulnerability, and responsiveness to others' needs, expanding ethical reflection beyond abstract rules to include lived experiences and interdependence.

Bibliography

"Aztec Moral Philosophy Didn't Expect Anyone to Be a Saint | Aeon Essays". n.d. *Aeon*. Accessed 10 July 2024. https://aeon.co/essays/aztec-moral-philosophy-didnt-expect-anyone-to-be-a-saint

Bentham, J. 1789. An Introduction to the Principles of Morals and Legislation. London: T. Payne.

Foot, Philippa. 1967. "The Problem of Abortion and the Doctrine of the Double Effect". *Oxford Review* 5: 5–15.

Gilligan, Carol. 1993. *In a Different Voice: Psychological Theory and Women's Development*. Vol. XXXVIII. Cambridge: Harvard University Press.

Kant, Immanuel, and Mary J. Gregor. 1997. *Critique of Practical Reason*. Cambridge Texts in the History of Philosophy. Cambridge: Cambridge University Press.

Kittay, Eva Feder. 2019. *Learning from My Daughter: Valuing Disabled Minds and Caring That Matters*. New York: Oxford University Press.

Korsgaard, Christine M. 1996. *The Sources of Normativity*. Edited by Onora O'Neill. Cambridge: Cambridge University Press.

MacIntyre, Alasdair C. 2007. *After Virtue: A Study in Moral Theory*. Vol. III. Notre Dame: University of Notre Dame Press.

Mill, John Stuart. 2001. *Utilitarianism: And the 1868 Speech on Capital Punishment*. Vol. II. Edited by George Sher. Indianapolis: Hackett Publishing Co.

Mills, Charles W. 2018. "Black Radical Kantianism". *Res Philosophica* 95 (1): 1–33. https://doi.org/10.11612/resphil.1622

Nozick, Robert. 2013. *Anarchy, State, and Utopia*. New York: Basic Books.

Puig de la Bellacasa, María. 2017. *Matters of Care: Speculative Ethics in More than Human Worlds*. Posthumanities 41. Minneapolis: University of Minnesota Press.

Purcell, S. 2018. "Aztec Moral Philosophy Didn't Expect Anyone to be a Saint". *Aeon*.

Regan, Tom. 2004. *The Case for Animal Rights*. Berkeley: University of California Press.

Singer, Peter. 1972. "Famine, Affluence, and Morality". *Philosophy & Public Affairs* 1 (3): 229–243.

———, and Yuval N. Harari. 2023. *Animal Liberation Now: The Definitive Classic Renewed*. New York: Harper.

Tronto, Joan C. 2013. *Caring Democracy: Markets, Equality, and Justice*. New York: New York University Press.

Väyrynen, Pekka. 2025. "Thick Ethical Concepts". In *The Stanford Encyclopedia of Philosophy* (Spring 2025 Edition), edited by Edward N. Zalta and Uri Nodelman. https://plato.stanford.edu/archives/spr2025/entries/thick-ethical-concepts/

3. Environmental Ethics

Introduction

Environmental ethics can be understood as a subfield in applied ethics. It seeks to include other-than-human life forms and the environment in moral discussions. While nature was very much present in philosophical discussions during the eighteenth and nineteenth centuries, environmental ethics only emerged as a distinct discipline during the 1960s and 1970s. The rapidly changing world and the environmental challenges caused by human activities have called for rethinking the relationship between humans and nature. This development came with the awareness that humans may not be the only morally worthy beings, hence the need to develop moral theories that could account for ecosystems, the environment, and other beings. Such a way of thinking may very well be 'stating the obvious' nowadays, but for a long time—at least in Western philosophy—little consideration had been given to other beings from a moral perspective. Caring for other-than-human animals and protecting them may seem intuitively right, and many have stood up against organizations and individuals like poachers who threaten them, but that has not always been the case.

Also, if many voices and currents of thought emerged over the years to give significant weight to non-human beings, a stark contrast remains. Not all views within environmental ethics argue for respecting non-human beings and nature on the same ground. There are two broad stances we need to distinguish here: anthropocentrism and ecocentrism. In anthropocentrism, although moral value is attached to non-human beings, humans remain the most important living entity. Ecocentrism, on the other hand, is a nature-centred approach that does not rely on a system of values that primarily applies to humans. In short, ecocentric views value nature for its own sake while anthropocentric views retain a value in nature for its instrumental use. It is crucial to understand that there is here an important ontological divide between both: ecocentrism rejects that there would be an existential split between the human world and the non-human world, while anthropocentrism on the contrary sees humans as separate from nature and superior to it. This is a rough sketch of the positions in debates around environmental ethics; it is important to bear in mind, however, that the distinctions between anthropocentric and non-anthropocentric views are complex, and that disagreements persist.

One of the initial tasks of environmental ethics was to come up with principles that could serve as a foundation for the field. Not surprisingly, in the same fashion as for moral theories, disagreements emerged between authors who proposed different principles to serve as guidelines. As one of the main impulses behind environmental ethics was to extend moral considerations to a broader circle of attention, one of the main questions environmental ethics needed to answer was 'whom and what' this circle should include. In other words, who or what is worthy of moral consideration? Philosophers from different moral strands (utilitarianism, deontology, virtue ethics) have proposed various answers to this challenging question and have advanced different criteria that could serve to draw a line between 'who is in' and 'who is out'. In this section, we look at some of the proposed major criteria.

One of the first philosophers to vigorously defend other-than-human interests was Peter Singer, whose book has served as a stepping stone for animal rights movements. In *Animal Liberation* (1975), Singer takes the position of a straightforward hedonic utilitarian (value is measured in terms of pleasure and pain). Hence, to be morally relevant, organisms must have the capacity to feel them. Because an organism has pleasure or pain, it has 'interests': it is interested in avoiding pain, or it is interested in sustaining or increasing pleasure. A tree, if we assume that it cannot feel pleasure or pain, thus has no interests and is outside of moral consideration. Singer's criterion is thus *sentiency*.

In its initial formulation, Singer's position suffered much criticism. He later expanded his hedonistic utilitarianism into a 'preference utilitarianism' (Singer, 1993). While the criterion remains, he distinguishes morally considerable beings on the basis of consciousness and self-consciousness. On the one hand, some conscious organisms feel pain and pleasure but have no self-awareness; they do not see themselves as persisting in the future, and hence, they have no preference to go on living. On the other hand, self-conscious organisms perceive themselves as individuals persisting through time, with desires and preferences for the future. Such preferences are, for Singer, morally relevant. Such a position entails that even if animals are morally worthy, some may be more worthy than others. All in all, sentiency remains the criterion determining who is in and who is out. From this perspective, all non-sentient organisms (such as plants, trees, and some animals) are morally irrelevant—except instrumentally, when they are a source of pleasure for sentient beings, for instance.

After Singer, Robin Attfield attempted to develop a more comprehensive consequentialist approach. For Attfield, moral worthiness is not based upon the capacity to experience pleasure or pain but instead on an ability to flourish (Attfield, 1987). Any organism that has the ability to flourish has an interest in doing so. All organisms are, therefore, morally considerable insofar as they do so; only inanimate objects are left morally inconsiderable. Attfield makes *flourishing* the criterion for inclusion or exclusion; however, it is important to note that, being a consequentialist, what matters for Attfield is the exercise of the basic capacities of a species. It is the state of flourishing that is valuable, not the individual organisms. While developing criteria for greater inclusion, philosophers rapidly noticed that, with a greater community of beings included, conflicts might emerge, and priority principles would be needed along with criteria. Later on, Attfield, in his book *A Theory of Value and Obligation* (1987), developed a two-pronged approach. One prong consists of a sliding scale of psychological complexity, with humans at the top and plants at the bottom. The other prong revolves around needs, interests, and wants. Basic and survival needs take precedence over mere preferences and wants, and the more complex organisms have priority over the simpler.

A third consequentialist approach was developed by Gary Varner. In *In Nature's Interests?*, he focuses on the *satisfaction of interests*, which are, according to him, possessed by all and only individual living things. With interests, an organism "has a welfare or good of its own that matters from a moral point of view" (Varner, 1998, p. 6). When it comes to priority, Varner sets up a hierarchy based on desires: some organisms (like animals) have desires, while others (like plants) do not. Thus, the interests of organisms that have desires outweigh those of organisms that do not. While his hierarchy still places humans—who even have higher desires that he calls ground projects (long-term desires that require satisfaction across a lifetime)—above all other organisms, it does not imply that humans should purely trump other beings' interests. For instance, taking into consideration that eating is necessary for humans to pursue their ground projects, his hierarchy allows for a certain granularity that makes eating plants (as non-desiring organisms) better than meat (as animals have desires).

Different critiques were made of these individualist consequentialists:

1. Identifying value with a certain form of experience (be it interests, pleasure, and pain through sentiency or flourishing) remains *anthropocentric*. In the end, these three views arbitrarily pick on a capacity possessed by humans and erect it as a paradigmatic quality, serving as a foundational criterion for their moral approach. Hence, without surprise, humans always end up at the top of the priority list since the whole logic behind it is to expand the circle based on a human capacity to include other beings.

2. The problem of *replaceability*. If what matters in the end is to maximize a state of utility (preferences, satisfaction of interests, or flourishing) deemed valuable, then it remains possible to discard, use, or sacrifice any organism if it leads to a better state of affairs.

3. The *practical implementation* of the proposed criteria might be difficult. How do we decide if a whale would be more sophisticated or complex than a bat or an ape, for instance? How are such decisions biased by human prejudice?

Let us leave these points of criticism aside for the moment to look at what criteria other approaches can offer. Albert Schweitzer wrote most of his work before the development of environmental ethics. Nevertheless, he profoundly influenced subsequent philosophers, most notably through his concept of *will-to-live*. According to Schweitzer, all living things have an impulse for self-realization that should be respected, and this is the basis for a universal concept of ethics: *Ehrfurcht vor dem Leben*, or Reverence for Life. "Ethics grow out of the same root as world- and life- affirmation, for ethics, too, are nothing but reverence for life. That is what gives me the fundamental principle of morality, namely, that good consists in maintaining, promoting, and enhancing life, and that destroying, injuring, and limiting life are evil" (Schweitzer, 1949, p. xviii). For Schweitzer, all beings with will-to-live are of equal value; humans are not in a position to judge those of other beings.

Picking up on this idea of respect for nature, Paul Taylor (2011) developed a biocentric approach with an Aristotelian background when he argued that all organisms are teleological centres for life. By that, he means that all organisms pursue some good of their own in their own way. As these organisms have a telos (goal/aim) which is vital for them, they have inherent worth. Realizing the difficulty of holding such an approach, Taylor subsequently developed four duties to the other-than-human natural world: non-maleficence, non-interference, fidelity, and restitutive justice. We will look at principlism in more detail in a later chapter. In the same fashion as consequentialist thinkers considered thus far, he also sets priority principles to resolve potential conflicts: self-defence, proportionality, minimum wrong, distributive justice, and restitutive justice. While these are complex issues we will not delve into, it is worth noting that through the simple examination of Taylor's view, there are different types of justice. We will examine these in the second part of this chapter.

Critiques were also raised against deontological approaches:

1. Both Taylor and Schweitzer have a concept of *restitution*, a form of ecological compensation according to which it would be possible to compensate for damage to or the death of individual organisms via good treatment of the same or different organisms. This holds some similarities with the replaceability issue consequentialist approaches faced. Furthermore, a restitution approach cannot be consistent with a deontological approach since wrongs cannot be summed up and compensated for as would be possible for consequentialists.

2. The moral worth given to all living beings is also problematic. Schweitzer and Taylor's views may be even more problematic than the consequentialist ones in practice. Asserting the equal value of all organisms, both fail to hold their positions consistently. Taylor, for instance, accepts medical treatment for humans during which millions of bacteria may die to save one life. Thus, despite what they advocate, both retain a certain form of hierarchy that could be problematic.

3. Other critics echo the same criticisms made of individualist consequentialist approaches. Here again, focusing on individual organisms does not allow us to ascribe value to wholes such as ecosystems or species—or, at least, it only allows it insofar as the individual members are valued. Equal value also leads to a certain (ontological) flattening by which diversity is of no value: a field of rare wildflowers is not worth more than a wheat field.

To address this criticism, some authors have attempted to develop holistic environmental approaches focusing on 'ecological wholes' (ecosystems, species, biosphere) rather than on organisms. Holistic approaches are mostly consequentialist, as they aim for the good of the whole. Holistic environmental ethics traces its roots back to Aldo Leopold's *A Sand County Almanac,* where he sets one of the most famous principles in the field: "A thing is right when it tends to preserve the integrity, stability, and beauty of the biotic community. It is wrong when it tends otherwise" (1949, p. 224). We can observe a striking contrast with the views laid out so far. First, the community is the focus of moral worth, and second, the quality used as a measurement criterion is intrinsic to this community. Environmental philosophers have further developed this holistic view after Leopold. For instance, Eric Katz (1983) suggests the ecological community's well-being as the primary ethical principle. While individual organisms have value as well, this is only secondary. Yet, the main holistic environmental defender is probably John Baird Callicott (1980), who strongly argues against any individualist approach. Firstly, he argues that all values are subjective and anthropocentric. Secondly, he accepts a form of sociobiology with the belief that ethical behaviour in human beings is instinctive and has been selected through evolution, with ethical responses by individuals within a biological community increasing survival. However, all living things have the same biological origins and form an interdependent community, which allows Callicott to emphasize the community rather than the individuals. We can quickly see some controversial conclusions from adopting such a view: for Callicott, some individuals may have to be sacrificed for the community's health. For instance, some pollinating bees are, in this view, more important than higher mammals such as humans, who sometimes are not only not vital to the community, but even a threat to it.

Again, it is no surprise that such views have triggered heavy criticism. Tom Regan, for instance, argues that holistic views allow to sacrifice individuals in the name of the stability, integrity, and beauty of the biotic community and suggests that these positions are nothing but environmental fascism (Regan, 1983, p. 362). Prioritizing the whole at the expense of the individuals is widely seen as ethically unacceptable. Callicott, like many, reviewed his positions in the light of such criticisms. He later tried to reconcile his views by introducing the idea of 'nested communities'. He argues that humans are intertwined with different moral communities which can be imagined as concentric circles, with ethical obligations diminishing towards the outer circles. He places humans in a core community at the centre; then there is a mixed community consisting of humans and domestic animals, and the wild biotic community on the outside. This review has different implications than his earlier positions, the two most important being the possibility of human concerns trumping those of other communities and the possibility of ascribing moral obligations to humans towards animals.

This brief account of different positions in environmental ethics shows us there is no silver bullet, no one-size-fits-all solution that extends moral considerations to non-human beings. Practically, they all face hurdles that challenge their practical application. While these issues continue, environmental ethical thinking has evolved to include other aspects, such as political ones. Beyond principles and criteria, some ways of thinking have become schools of thought. We will consider two of the most important ones, deep ecology and ecofeminism, to see how environmental thinking has changed over time to include broader considerations and solve environmental challenges while attempting to reconcile different standpoints.

Deep ecology

Norwegian philosopher Arne Naess coined the term 'deep ecology' in *The Shallow and the Deep, Long Range Ecology Movements: A Summary* (1973). In his article, he distinguished two strands of ecological movements: the shallow, chiefly concerned with pollution and resource depletion; and the deep, characterized metaphysically, ethically, and politically. On metaphysical grounds, deep ecology rejects the idea that humans are separate from their environment and that all things have complex interrelatedness. Ethically, deep ecology recognizes a biocentric equality principle—that there is equal value to all living things—but acknowledges that practically, realistically, some exploitation and killing cannot be avoided. Deep ecology also requires political action to put the principles into practice while favouring diversity and decentralization. Like other authors supporting biocentric equality, deep ecology faced criticism, which led Naess to rearticulate his ideas in the Deep Ecology Platform, a series of principles serving as a foundation for the movement (Palmer, 2003).

Ecofeminism

Deep ecology has inspired various political and environmental movements such as Earth First! Nevertheless, deep ecology faced intense criticism. It did not solve all the difficulties with affirming intrinsic value and egalitarianism. Ecofeminism has strongly argued against some principles of deep ecology. Ecofeminist philosophy traces its roots back to deep ecology until it diverged from it in the 1980s and 1990s (Warren, 2015). Essentially, ecofeminists came to disagree with two basic principles of deep ecology. Firstly, with the principle that all organisms have equal and inherent value—if this principle successfully rejects anthropocentrism, it fail to account for anthropocentrism being androcentric. Ecofeminism sprung from the idea that the domination and oppression of nature is somewhat akin to the oppression of women by men. This twin oppression is understood in different ways by different ecofeminists, and so are the methods to remediate it. Nevertheless, there is a consensus among ecofeminists that, due to the connection between these axes of oppression, ecological accounts must be informed by a feminist perspective and vice versa.

They have also argued against the principle of self-realization, according to which the human self is actualized only if it is merged with nature. According to the philosopher and ecofeminist Val Plumwood: for deep ecology, the key problem in the relationships between humans and nature comes from their separation (1991). The problem with this account, according to Plumwood, is that it allows the self to operate on the fuel of self-interest despite the potential for a wider set of concerns. Other-than-humans, in this account, have moral status only to the extent that they can be incorporated into the self, which would deny their differences. As such, deep ecology may lead to the denial of particular meanings as well as deep and particularistic attachments to places.

Aside from these criticisms, ecofeminist philosophy retains shared views with deep ecology and advocates for a 'spiritual' identification with nature in reverence for life processes, without seeking its utility to humans (Birkeland, 2010). Through more than critical analyses of dominant paradigms, ecofeminist philosophy offers interesting alternatives in the account it puts forward based on the importance of relationships and care (Plumwood, 1991), the emotional force of kinship, or closeness to another (Gruen, 1993). In other words, by emphasizing that humans are entangled in nature and do not work in isolation. Everything is interconnected and cannot be understood in isolation apart from context or ecological niche (Kheel, 1993).

Vandana Shiva, a scientist, environmental activist, and ecofeminist, criticizes the industrial and technocratic approaches to agriculture and science that reduce nature—and often women—to passive resources to be exploited. In contrast, she proposes Earth Democracy, a vision grounded in the interconnectedness of all life, where care for the planet is inseparable from social justice (Shiva, 2005).

Ecofeminism—drawing on other strands of feminist philosophy—maintains that there are no value-neutral, universally applicable, unbiased points of view. All ethical standpoints are the products of different worldviews, contexts, and places. Care should serve as the basis for all relationships between humans and the other-than-human world. It makes relationships (which are essential in defining who one is), rather than individual organisms, the valuable qualities central to ethical decision-making.

Despite how appealing and reconciliatory ecofeminist philosophy may be, it has faced its share of criticism. For one, the contextual approach may imply that there are no universals or absolutes. Another criticism is aimed at care as the foundation of environmental ethics: does it make sense to even talk about relationships or care with non-living entities (Palmer, 2003)? Does it make sense to talk to and care for a stone? How could one do so for something that supposedly has no inner state and cannot be made better or worse?

This short glimpse at deep ecology and ecofeminist philosophy shows how diverse views about environmental ethics can be. It also shows how complex the debates are, fuelled with heavy criticism from all strands. Nevertheless, environmental ethics has grown into a major field in ethics and, without a doubt, will keep growing. Ecological challenges have drawn attention to ethical insights in considering other-than-humans, ecosystems, and the biosphere itself. The task of environmental ethics is to elaborate on which answers are appropriate and what should be done in the future to tackle these challenges. With wilderness declining, urbanization growing, new forms of pollution, and human displacement, debates will expand to consider wider kinds of environments and the ethical issues they raise.

A mini philosophy of microbes

Although largely ignored by the discipline, microbes are philosophically interesting (O'Malley, 2014). We will focus on two reasons why this is the case.

Some philosophers of biology have questioned the fact that both evolutionary biologists and philosophers have mostly taken interest in organisms that are close to human beings. Microbes such as bacteria challenge existing concepts of taxonomy. Present-day concepts of taxonomy are based on sexual reproduction, genetic relatedness, and vertical gene transfer. And although different 'species' of bacteria have been identified, methods of classification based on genetic relatedness fail, as bacteria that are lumped together under one species may have genomes that are much more different than in mammalian species. Moreover, bacteria can exchange genes through a process called 'horizontal gene transfer' with other bacteria that may not strictly belong to the same species. Sexual reproduction and vertical gene transfer has long been considered the default and has been used as a basis to pinpoint fundamental biological laws of behaviour. However, as bacteria and horizontal gene transfer precede any other form of life, the possibilities this presents for different ways of looking at life cannot be understated. Bacteria can also teach us about philosophers' favourite question; 'what is a human being?' Although we often think of microbes in terms of good and bad, findings in microbiology show that our relationship with them is more complicated. For one, bacteria and even viruses are rarely purely 'good' or 'pathological': whether they affect human and more-than-human health depends on the specific relation and context, such as where they are located in the body. Moreover, discoveries related to the microbiome-gut-brain axis demonstrate how the microbes in our gut are linked with behaviour and even personality. This challenges 'atomistic' views of human beings as independent standalone organisms.

Environmental justice

Environmental justice, health, and antibiotic resistance

Antibiotics resistance

As we saw, the rapid changes in our world have presented us with many environmental ethical issues. Beyond questions of which (living) entities matter and on what ground, the interconnectedness of phenomena and their measurable (as well as unmeasurable) impacts have raised questions on a global scale about environmental justice, the most salient probably being climate change and its consequences. Nevertheless, there are many complicated issues of global environmental justice.

A growing, pressing issue— intertwining health, environmental, and justice dimensions—is the use of antibiotics.

The world is now confronted with an unprecedented crisis, facing the risk of a post-antibiotic era where infections that were treatable for decades will kill again. Firstly, antibiotics do not make distinctions between which microbes are responsible for a specific infection and which are not, so their widespread use has raised concerns about the disappearance of microbes which can contribute positively to the health of larger organisms (Gilbert and Epel, 2015, p. 123). As a result, human physiology will change, and so will human health. Additionally, the use of antibiotics leads to antibiotic resistance (ABR). ABR threatens to undermine global public health by erasing decades of progress in medicine, food security, and public health (Laxminarayan et al., 2016). By 2050, 300 million people could die as a result of ABR, which would also have dire economic consequences, with estimated financial loss of up to $100 trillion (O'Neill, 2014, p. 6).

Antibiotics do not simply vanish after usage but are released into the environment in many ways. The urines and feces of users—which contain high amounts of the active substances comprising many antibiotics—are directly released into the environment in countries lacking proper sewage infrastructures. In some countries, treatment plants retain a high concentration of antibiotics which microbial communities can be exposed to (Larsson, 2014). Antibiotics are also inserted in the environment during the production processes of active pharmaceutical ingredients and through unused discarded medicines (Larsson, 2014). The release of antibiotics into the environment is problematic because it plays a preponderant role in ABR. Firstly, it influences the emergence and evolution of pathogens. The release of antibiotics increases environmental pressure on bacteria to develop more resistance in response. The environment turns into an arena providing conditions where the added pressure can increase the available pool of resistance genes and selection for bacteria to acquire resistance through horizontal gene transfer (Larsson, 2014). Secondly, the environment plays a role in the transmission and dissemination of resistant bacteria (Bengtsson-Palme et al., 2018), which can happen through contaminated water spreading bacterial pathogens due to human's extensive traveling and through the transportation of goods and food (Larsson et al., 2018). This has led to the emergence of the One Health perspective, which considers human and other-than-human animal health to be interconnected with the environment. Yet, so far, no research has shown the exact direct and indirect impacts on health resulting from environmental exposure to antibiotic-resistant bacteria (Wuijts et al., 2017).

From the perspective of responsibility regarding environmental pollution, ABR raises many ethical challenges. Some are directly linked to health issues and exacerbate them, and others are linked to questions of global justice. Four different sets of ethical problems can be distinguished (Littmann et al., 2015). (1) ABR impacts ethical challenges in infectious disease control where patient autonomy needs to be balanced with the protection of others. (2) The second set of problems pertains to animal ethics and the use of veterinary antibiotics, which represents more than half of the world's antibiotic consumption. Antibiotics are used not only to treat animals but also as a growth stimulant. While banning the non-therapeutic use of antibiotics in the veterinary sector is ethically quite uncontroversial, limiting their use to treat animals raises questions of animal welfare rarely addressed (Littmann et al., 2015). (3) ABR raises challenges of distributive justice for the fair allocation of antibiotics. If it is urgent to reduce antibiotic consumption, there are still many people who die due to a lack of access to high-quality antibiotics. It is necessary to reduce the excessive use of antibiotics in some areas while ensuring their access in other areas (Littmann and Viens, 2015). Addressing the ABR issue collectively, on a global scale, means determining who is responsible for reducing antibiotic use and to what extent, who needs assistance to reduce their use, and how we can still benefit from this resource. (4) Finally, ABR raises inter- as well as intra-generational ethical challenges of distributive justice in that current generations' interests in antibiotics conflict with the interest of future generations who will have to bear the consequences of antibiotics with reduced efficacy (Littmann et al., 2015). Thus, ABR shares many similarities with climate change as a multisectoral problem, making it difficult to address. In short, it is another pressing 'perfect moral storm', which is why it is relevant to consider.

The following chapters about health care ethics and animal ethics will flesh out issues (1) and (2) and allow you to better grasp all the dimensions and complexity of ABR. But first, let us return to the other issues. Namely, that ABR raises questions of justice and more specifically distributive justice.

What is (environmental) justice?

Justice takes on different meanings in different practical contexts, and to understand it fully, we have to grapple with this diversity. But first, let's have a look at what 'justice' means. Philosophically, some of the most ancient theories of justice came from the Greeks. For them, justice was, first and foremost, a virtue: the virtue of the soul (in action) (Miller, 2023). Aristotle and Plato both conceived justice as goodness and tied it to an ideal of perfection in human relationships. Although this may sound abstruse, we can already draw some initial insights from their perspectives: justice has something to do with 'being good' and is held as having the highest value. Also, from this angle, justice has more of a moral meaning than a legal one. Perhaps the first established core definition of justice traces back to the Institutes of Justinian, from the sixth century AD. In Roman Law, justice was defined as "the constant and perpetual will to render to each his due" (Miller, 2023). This definition encapsulates several dimensions of justice that still make sense nowadays. In other words, justice looks at how humans are treated to ensure that each person gets what they deserve equitably and consistently through time.

Justice is complex. Not only does it encompass several dimensions, it also implies discussions from different angles in different frames of reference. In short, depending on the context, justice can mean or refer to many different conceptions. Environmental justice mostly refers to a distributive kind of justice. Justice can be employed as a distributive principle (Lamont and Favor, 2017): when there are resources or goods that need to be distributed and several people have a claim over them, distributive justice aims at ensuring that the repartition is fair. Fairness can be understood in various ways; hence, a just distribution could take various forms. How do we ensure that an apple pie is shared fairly if it has to be cut into slices for several people? A fair distribution means that everyone should get a slice of the same size. Others could argue that the hungriest people should get bigger slices. Another possible factor to consider is that the people who harvested the apples and baked the pie have put some effort into it and should be rewarded for this effort. It is also possible to argue that the people who like apples the most should get bigger slices because they are the ones who would get the most pleasure from eating the pie.

Hence, environmental justice is a type of distributive justice chiefly concerned with the distribution of risks and benefits linked to the environment. So, it deals with questions about who should bear the costs of pollution, who should have access to different kinds of resources, and so on. Now, it should be more clear why ABR is a case of distributive justice. First of all, it concerns a fundamental aspect of human life: health and access to care. On a global scale, the inequalities are well-known and widespread, life expectancies differ greatly from one country to another, and the availability of antibiotics is much more critical for populations in dire need and/or populations without access to alternative treatments. Types and concentrations of antibiotics vary greatly from one country to another (Hanna, et al., 2018). The current state of affairs is a result of unequal development and access to high-quality antibiotics—some low-income countries never even fully entered the antibiotic age (Littmann et al., 2015) and are disproportionally affected by ABR.

But it gets more complicated, like climate change; ABR raises inter- and intra-generational ethical challenges of distributive justice where the interests of different generations compete. Most discussions consider forward-looking perspectives by including future generations, but we can also look backward to integrate remediation for past harms. Then, justice is not only taken as a distributive principle but as a remedial principle: the justice we talk about when a wrong was committed, which seeks to restore the victim's state of affairs had the wrong not occurred (Miller, 2023). In the case of ABR, while countries with early access to antibiotics had unlimited use of them until ABR was discovered, other countries—which now account for a large portion of antibiotic consumption (Laxminarayan et al., 2015)—risk being denied the same use now that restrictions are to be applied all around the world to safeguard the efficiency of antibiotics. This could give ground for compensation on two aspects: developing countries should be compensated for the missed opportunities antibiotics restriction would impose on them, and developed countries should also compensate for their past overuse that has jeopardized antibiotic efficiency. Compensation also makes sense if we look at it from the perspective of production processes. Pharmaceutical companies have developed antibiotics mostly for the benefit of high-income countries, but the antibiotics themselves are often produced in other countries, especially in China and India (Hanna et al., 2018). The production of antibiotics polluted the environments of these countries, increasing the presence of antibiotic-resistant genes in the areas (Ashbolt et al., 2013). For instance, in the Ganges valley, water used for drinking and recreational purposes is now contaminated by resistant bacteria.

Through these short considerations, which only barely cover the justice aspects of ABR challenges, we can see how environmental ethics and justice issues can quickly grow in complexity and that justice or 'being just' is not as straightforward as one may think.

Environmental justice: the case of Ecuador

The Yasuni case

In the east of Ecuador in Latin America, the Yasuni National Park encompasses a section of the Amazon rainforest. Recognized as one of the most biodiverse regions in the world, it is also home to several Indigenous Amazonian tribes. In this rainforest, an oil field named Ishpingo-Tambococha-Tiputini (ITT) was discovered in the early 2000s, containing approximately 846 million barrels of crude oil—around 20% of Ecuador's proven oil reserves.

> **Exercise:** Watch the video clip on the oil drillings in Yasuni National Park (4.5 mins).
>
> https://www.youtube.com/watch?v=c07Z1ZexT7E
>
> Think about and list different arguments for and against the extraction of oil in Yasuni National Park.

What is of value? Anthropocentrism and ecocentrism

Whether or not to drill for oil is the question in this case. In listing arguments for and against, one might say that another round of large-scale deforestation in the Amazon is bad. But why? Is it because we cherish the value of the individual trees, forests, and the ecosystems they support? Or is cutting down rainforests bad since it deprives humans of so-called 'ecosystem services' which might buffer climate change or provide a reservoir of undiscovered medicines? Of course, these kinds of arguments can be combined, yet this question of 'what is of value?' has been essential in environmental ethics.

The Yasuni-ITT initiative

Let us return to our case in Ecuador. To prevent oil exploitation, Indigenous peoples and environmental movements devised an alternative proposal that the Ecuadorian government eventually picked up. The proposal, called the *Yasuni-ITT initiative*, suggested a permanent ban on oil production inside the Ishpingo-Tambococha-Tiputini oil field. In exchange, the Ecuadorian president asked the international community at a general assembly of the United Nations to contribute to a fund worth 50% of the value of the Yasuni oil reserves, or $3.6 billion. By preventing the drilling, the Yasuni-ITT initiative sought to conserve the region's biodiversity and protect the Indigenous peoples currently living in voluntary isolation inside the Yasuni National Park (Einhorn, Andreoni, and Schaff, 2023). It also sought to avoid the emission of significant quantities of CO_2 caused by burning the oil, and to begin a transition to a sustainable economy in Ecuador, using the funds to create jobs in sectors such as renewable energy.

> Think about the distribution of benefits, harms, and responsibilities in either of two scenarios:
> 1. The extraction and selling of the entire oil reserve.
> 2. The Yasuni-ITT initiative to request financial compensation from the international community to leave the oil in the ground.
>
> Which people/groups benefit from either option? Who is harmed? Who is responsible?

Buen Vivir

Buen Vivir, or Vivir Bien, is a term used in Latin American countries like Ecuador and Bolivia to describe a moral and political worldview of the good life in a broad sense (Gudynas, 2011). Buen Vivir is a Spanish translation of the Kichwa (or Quechua) term *sumak kawsay* and similar terms, which refer to a fullness of life in a community, together with other persons and nature, situated in a specific territory. The Buen Vivir worldview is built on the traditions and knowledge of Indigenous peoples living in the Andes and Amazon regions for centuries (Gudynas, 2011). Thus, the immediate environment is seen as part of the community people belong to, which positions Buen Vivir as an ecocentrist worldview.

Since the early 2000s, Indigenous ideas started to be combined with a present-day critique, leading to the Spanish phrase 'Buen Vivir'. This critique mainly targeted notions of 'development' and 'extractivism', which dominated Latin American countries over the past five decades (Fatheuer, 2011). Extractivism refers to the ongoing, large-scale extraction of material resources (gold, lithium, oil, etc.) in countries in the Global South with the often unfulfilled promise that profits will be used to 'develop' the country in social and economic terms. The Yasuni case is a clear example which fuelled Buen Vivir's expansion in Ecuador.

Rights of nature

Proponents of Buen Vivir stress that their ideas are always evolving, but some main concepts can be distilled. For example, they favour an economy that is directly in service of people and their environment, grounded on principles of reciprocity, sufficiency, and solidarity, in clear contrast to the extractivist idea that the environment needs to be capitalized on before people's lives can be improved in a later stage.

A key breakthrough of the Buen Vivir worldview is the explicit inclusion of the term and some of its central ideas into Bolivia's and Ecuador's revised constitutions. Since 2008, Ecuador's constitution has contained several articles on 'the rights of Nature'. This is a unique and remarkable move, since constitutions generally describe the rights of people and their right to a clean environment. In Ecuador, 'Nature' actually holds rights in itself. Article 71, for example, says, "Nature, or Pacha Mama, where life is reproduced and occurs, has the right to integral respect for its existence and the maintenance and regeneration of its life cycles, structure, functions, and evolutionary processes". Thus, the environment's intrinsic value—which ecocentrists argue in support of—was translated and anchored in a juridical document. Also, the direct moral obligation to protect 'Nature' is embedded in the constitution: "The State shall apply preventive and restrictive measures on activities that might lead to the extinction of species, the destruction of ecosystems and the permanent alteration of natural cycles" (Article 73).

Importantly, these Rights of Nature do not stipulate an obligation to protect every *individual* plant, animal, river, etc., in themselves. Instead, it takes a system theory's perspective, speaking of life cycles, ecosystems, and evolutionary processes that need to be valued and protected. In other words, Buen Vivir does not call for non-interference with the environment or conservation of pristine nature. However, it does call for a human-environmental relationship based on reciprocity and interdependency.

Conclusion

In this chapter, we have seen how the field of environmental ethics has evolved, from its inception in the wake of pollution cases that emerged in the 1960s to its most recent developments, taking the form of full-fledged theories. The first approaches to environmental ethics attempted to broaden the circle of morally worthy beings beyond humans by using different inclusion criteria, such as sentience. Nevertheless, these approaches largely remained anthropocentric and faced criticism. The field of environmental ethics subsequently evolved with all-encompassing theories, such as deep ecology and ecofeminism, that linked environmental issues with political and social matters. The second part of this chapter discussed what environmental justice is through the lens of antibiotic resistance and the Yasuni Park case in Ecuador. These examples highlighted the complexity of being 'just' from an environmental perspective, due to the scale of the issues it covers and the complex interrelations and interactions it involves.

Bibliography

Ashbolt, Nicholas J., Alejandro Amézquita, Thomas Backhaus, Peter Borriello, Kristian K. Brandt, Peter Collignon, Anja Coors, Rita Finley, William H. Gaze, Thomas Heberer, John R. Lawrence, D.G. Joakim Larsson, Scott A. McEwen, James J. Ryan, Jens Schönfeld, Peter Silley, Jason R. Snape, Christel Van den Eede, and Edward Topp. 2013. "Human Health Risk Assessment (HHRA) for Environmental Development and Transfer of Antibiotic Resistance". *Environmental Health Perspectives* 121: 993–1001. https://doi.org/10.1289/ehp.1206316

Attfield, Robin. 1987. *A Theory of Value and Obligation*. London: Routledge.

Bengtsson-Palme, Johan, Erik Kristiansson, and D. G. Joakim Larsson. 2018. "Environmental factors influencing the development and spread of antibiotic resistance". *FEMS Microbiology Reviews* 42 (1). https://doi.org/10.1093/femsre/fux053

Birkeland, Janis. 2010. "Ecofeminism: Linking Theory and Practice". In *Ecofeminism Women Animals Nature*, edited by Greta Gaard, pp. 13–59. Philadelphia: Temple University Press.

Callicott, J. Baird. 1980. "Animal Liberation: A Triangular Affair". *Environmental Ethics* 2 (4): 311–38. https://doi.org/10.5840/enviroethics19802424.

Einhorn, Catrin, Manuela Andreoni, and Erin Schaff. 2023. "Ecuador Tried to Curb Drilling and Protect the Amazon. The Opposite Happened". *The New York Times*, 14 January 2023. https://www.nytimes.com/2023/01/14/climate/ecuador-drilling-oil-amazon.html.

Fatheuer, Thomas. 2011. *Buen Vivir: A Brief Introduction to Latin America's New Concepts for the Good Life and the Rights of Nature*. Berlin: Heinrich Böll Foundation.

Gilbert, Scott F., and David Epel. 2015. *Ecological Developmental Biology: The Environmental Regulation of Development, Health, and Evolution*. 2. Oxford: Oxford University Press.

Graham, Kevin M. 2011. "Non-Ideal Moral Theory". In *Encyclopedia of Global Justice*, edited by Deen K. Chatterjee, pp. 758–60. Dordrecht: Springer Netherlands. https://doi.org/10.1007/978-1-4020-9160-5_124.

Gruen, Lori. 1993. "Dismantling Oppression: An Analysis of the Connection Between Women and Animals". In *Ecofeminism Women Animals Nature*, edited by Greta Gaard, pp. 60–90. Philadelphia: Temple University Press.

Gudynas, Eduardo. 2011. "Buen Vivir: Today's Tomorrow". *Development* 54 (4): 441–47. https://doi.org/10.1057/dev.2011.86.

Gustafsson, Martin. 2004. "On Rawls's Distinction between Perfect and Imperfect Procedural Justice". *Philosophy of the Social Sciences* 34 (2): 300–5. https://doi.org/10.1177/0048393104264925.

Hanna, Nada, Pan Sun, Qiang Sun, Xuewen Li, Xiwei Yang, Xiang Ji, Huiyun Zou, Jakob Ottoson, Lennart E. Nilsson, Björn Berglund, Oliver James Dyar, Ashok J. Tamhankar, and Cecilia Stålsby Lundborg. 2018. "Presence of antibiotic residues in various environmental compartments of Shandong province in eastern China: Its potential for resistance development and ecological and human risk". *Environmental International* 114: 131–142. https://doi.org/10.1016/j.envint.2018.02.003

Katz, Eric. 1983. "Is There a Place for Animals in the Moral Consideration of Nature?" *Ethics and Animals* 4 (3). https://doi.org/10.15368/ea.1983v4n3.1

Kheel, Marti. 1993. "From Heroic to Holistic Ethics: The Ecofeminist Challenge". In *Ecofeminism Women, Animals, Nature*, edited by Greta Gaard, pp. 243–271. Philadelphia: Temple University Press.

Kortetmäki, Teea. 2013. "Anthropocentrism versus Ecocentrism Revisited: Theoretical Confusions and Practical Conclusions". *SATS* 14 (1). https://doi.org/10.1515/sats-2013-0002

Lamont, Julian, and Christi Favor. 2017. "Distributive Justice". In *The Stanford Encyclopedia of Philosophy*, edited by Edward N. Zalta. Stanford: Stanford University. https://plato.stanford.edu/archives/win2017/entries/justice-distributive/

Larsson, D. G. J. 2014. "Antibiotics in the environment". *Upsala Journal of Medical Sciences* 119(2): 108–112. https://doi.org/10.3109/03009734.2014.896438

——, Antoine Andremont, Johan Bengtsson-Palme, Kristian Koefoed Brandt, Ana Maria de Roda Husman, Patriq Fagerstedt, Jerker Fick, Carl-Fredrik Flach, William H. Gaze, Makoto Kuroda, Kristian Kvint, Ramanan Laxminarayan, Celia M. Manaia, Kaare Magne Nielsen, Laura Plant, Marie-Cécile Ploy, Carlos Segovia, Pascal Simonet, Kornelia Smalla, Jason Snape, Edward Topp, Arjon J. van Hengel, David W. Verner-Jeffreys, Marko P. J. Virta, Elizabeth M. Wellington, and Ann-Sofie Wernersson. 2018. "Critical knowledge gaps and research needs related to the environmental dimensions of antibiotic resistance". *Environment International* 117: 132–138. https://doi.org/10.1016/j.envint.2018.04.041

Laxminarayan, Ramanan, Precious Matsoso, Suraj Pant, Charles Brower, John-Arne Røttingen, Keith Klugman, and Sally Davies. (2016). "Access to effective antimicrobials: A worldwide challenge". *Lancet* 387 (10014): 168–175. https://doi.org/10.1016/S0140-6736(15)00474-2

Leopold, Aldo. 1949. *A Sand County Almanac*. New York: Oxford University Press.

Littmann, Jasper, Alena Buyx, and Otto Cars. 2015. "Antibiotic resistance: An ethical challenge". *International Journal of Antimicrobial Agents* 46 (4): 359–361. https://doi.org/10.1016/j.ijantimicag.2015.06.010

——, and A. M. Viens. 2015. "The Ethical Significance of Antimicrobial Resistance". *Public Health Ethics* 8 (3): 209–224. https://doi.org/10.1093/phe/phv025

Miller, David. 2023. "Justice". In *The Stanford Encyclopedia of Philosophy*, edited by Edward N. Zalta and Uri Nodelman. Stanford: Stanford University. https://plato.stanford.edu/archives/fall2023/entries/justice/.

Naess, Arne. 1973. "The Shallow and the Deep, Long-Range Ecology Movement. A Summary". *Inquiry* 16: 95–100.

O'Malley, M. 2014. *Philosophy of Microbiology*. Cambridge: Cambridge University Press.

O'Neill, J. 2014. "Antimicrobial Resistance: Tackling a Crisis for the Health and Wealth of Nations". *Review on Antimicrobial Resistance*.

Palmer, Clare. 2002. "An Overview of Environmental Ethics". In *Environmental Ethics: An Anthology*, edited by Andrew Light and Holmes Rolston III, 1st edition. Malden: Wiley-Blackwell.

Plumwood, Val. 1991. "Nature, Self, and Gender: Feminism, Environmental Philosophy, and the Critique of Rationalism". *Hypatia* 6 (1): 3–27.

Rawls, John. 1971. *A Theory of Justice: Original Edition*. Cambridge: Harvard University Press. https://doi.org/10.2307/j.ctvjf9z6v.

Regan, Tom. 1983. *The Case for Animal Rights*. Los Angeles and Berkeley: University of California Press.

Schweitzer, Albert. 1949. *Civilization and Ethics*. London: Adam & Charles Black.

Singer, Peter. 1975. *Animal Liberation*. London: HarperCollins.

——— 1993. *Practical Ethics*. 2nd edition. Cambridge: Cambridge University Press.

Shiva, Vandana. *Earth Democracy: Justice, Sustainability, and Peace*. Cambridge, MA: South End Press, 2005.

Taylor, Paul W. 2011. *Respect for Nature: A Theory of Environmental Ethics - 25th Anniversary Edition*. Princeton: Princeton University Press.

Varner, Gary. 1998. In Nature's Interests? Interests, Animal Rights, and Environmental Ethics. Oxford: Oxford University Press.

Warren, Karen J. 2015. "Feminist Environmental Philosophy". In *The Stanford Encyclopedia of Philosophy*, edited by Edward N. Zalta. Stanford: Stanford University. https://plato.stanford.edu/archives/sum2015/entries/feminism-environmental/.

———, and Jim Cheney. 2003. "Ecological Feminism and Ecosystem Ecology". In *Environmental Ethics: An Anthology*, edited by Andrew Light and Holmes Rolston III, 1st edition. Malden: Wiley-Blackwell.

Wenzel, Michael, Tyler G. Okimoto, Norman T. Feather, and Michael J. Platow. 2008. "Retributive and Restorative Justice". *Law and Human Behavior* 32 (5): 375–89.

Wuijts, Susanne, Harold H. J. L. van den Berg, Jennifer Miller, Lydia Abebe, Mark Sobsey, Antoine Andremont, Kate O. Medlicott, Mark W. J. van Passel, and Ana Maria de Roda Husman. 2017. "Towards a research agenda for water, sanitation and antimicrobial resistance". *Journal of Water and Health* 15 (2): 175–184. https://doi.org/10.2166/wh.2017.124

4. Health Care Ethics

Introduction

Bioethics is most commonly associated with questions on health and disease. Matters such as end-of-life care, abortion, or preimplantation genetic testing are often what first springs to mind when one hears about 'bioethics'. While these are important topics to be discussed on a societal scale, the clinical encounter and the patient-physician relationship similarly introduce important moral questions on the interpersonal level of healthcare. Illness and disease usually result in the patient presenting in a vulnerable state. Both practically and institutionally, we put our trust in healthcare professionals to provide the necessary care per our values. These conditions come with significant risks of exploitation.

Physicians, nurses, and other healthcare workers often face difficult ethical decisions on the other side of the sick bed. For example, how should they handle unresponsive patients, or patients whose treatment options have been exhausted? What about patients making 'irrational' decisions? How should care be prioritized under conditions of resource scarcity? These are but some of the moral dilemmas that arise in the clinic.

As the recent global pandemic acutely brought to light, matters of public health can also be relevant to the lives of (healthy) individuals. Debates on allocating hospital beds, vaccines and other therapeutic resources, lockdown measures, and vaccine mandates—while challenging for policymakers—also offer puzzling cases for ethicists.

Biomedical researchers, too, are often confronted with complex ethical dilemmas. Clinical trials require (healthy) participants and patients to voluntarily and selflessly put themselves at risk for research on disease mechanisms, novel treatments, or diagnostic tools. Even if the outcomes of such research (can) constitute social goods, biomedical research also raises questions on how to ensure that research participants are not exploited, who should benefit from such research, and which types of research should be prioritized.

In sum, all parts of biomedical science and healthcare introduce significant moral challenges and trade-offs. In this chapter, we will be looking at several topical issues in healthcare ethics. But before we get into some of the ethical frameworks that researchers and physicians rely on, it is worth pointing out that healthcare ethics is a highly interdisciplinary field. In addition to philosophers, biomedical researchers and health care professionals—including physicians, primary care workers, and nurses—also often sit on committees for medical ethics. Outside of academia, various forms of activism are increasingly and significantly influencing bioethical discourse. As is to be expected, these different standpoints often offer radically different appraisals of traditional questions in bioethics—we will encounter some examples further on. A proper, flexible ethical toolkit is necessary to appreciate and accommodate such differing viewpoints into meaningful ethical decisions and guidelines.

Doing (medical) ethics

In chapter 2, we discussed how committing yourself to one moral theory, such as utilitarianism or deontology, is often ill-advised in complex, real-life cases. This is intuitively clear in ethical reflections on medical and public health questions, where our moral intuitions can take us in wildly different directions based on the specifics of the case. Recall, for example, the utilitarian surgeon who could save five lives by harvesting the organs of a single patient. Most would agree that this constitutes a clear transgression of the individual's autonomy—a deontological principle we generally consider important in medical decisions.

Compare this with public health cases where utilitarian reasoning sometimes may provide a sensible answer to an ongoing crisis. Think of measles eradication due to vaccine mandates, or the lockdown measures during the COVID-19 pandemic. In those instances, concerns over public health (i.e. protecting vulnerable populations or avoiding the collapse of our healthcare system) were considered sufficient to support the restriction of individual liberties.

Ethical questions generally do not allow for clear-cut answers. This is immediately evident when considering medical decisions. Patients may evaluate a particular medical result very differently or care about very different aspects of health and well-being. While it may seem clear that smoking increases the risk of lung cancer, for some, smoking a cigarette might be a welcome distraction from a stressful, busy day and might thus contribute to well-being. Or consider the following: for some prospective parents, trisomy 21 (a third copy of chromosome 21, often leading to developmental delays or intellectual disability in the form of Down's syndrome) might be a valid reason to opt for pregnancy termination; while for others, it might not be. For some, breast amputation and reconstruction is a reasonable response to an identified or assumed (genetic) risk for breast cancer, while others approach such situations differently. Even though we sometimes think we can conclusively determine what is in the patient's 'best interest', this might not always be clear. In sum, medical decisions are often complex and involve a variety of actors (physicians, patients and their families/community, etc.) all operating according to different value frameworks. Bioethical reflection needs to be able to deal with these (conflicting) values.

Ethical pluralism

As ethicists, we must have sufficiently sensitive tools to draw out these different features and viewpoints in a particular case. As such, most ethicists rely on multiple moral principles to come to a sensible conclusion. This stance is often referred to as *ethical pluralism*. In an earlier chapter, we saw how, according to Kant, certain maxims and specific universal laws are absolute: you have a *duty* to tell the truth even if a murderer is knocking at the front door. In contrast to the monist view (only using one ethical theory to assess a case), in applied ethics—of which medical ethics is a more specialized field—ethicists have devised multiple, specific principles to help make an ethical decision.

William David Ross (1877–1971) proposed that such principles are *prima facie* (Ross, 2002). These principles—which we will discuss in more detail later in this chapter—seem valid as duties *at first glance*; but when applied to concrete situations, they often must be weighed against one another. According to Ross, ethics is about how to act in *specific* situations. It follows that our duties might also depend on particular situations and relations between different actors. What might seem, at first glance, to be an ethical duty may be superseded by other aspects relevant to the case. Moreover, what might be a duty for one actor—for example, a doctor—may not be a duty for a patient. Therefore, rather than providing clear-cut answers to difficult questions, ethical reflection is complex and indecisive as to *whether we can* draw the correct conclusions. In conditions of uncertainty—as in most ethical cases—it is thus crucial that we (can) assess the complexities of a case before we come to decisions.

To facilitate decision-making and to consider all relevant aspects of a case, Ross proposed seven *prima facie* duties or principles:

- Duty of fidelity (promise-keeping)
- Duty of reparation (making up for prior wrongful acts)
- Duty of gratitude (being grateful for others' acts of kindness)
- Duty of justice (being fair)
- Duty of beneficence (benefiting or helping others)
- Duty of self-improvement (education or practice)
- Duty of non-maleficence (not harming others)

Principlism: the basic idea

In the wake of Ross' list, different subdisciplines of applied ethics have thought of specific principles or duties that capture important values relevant to their domain. Medical ethicists Tom Beauchamp and James Childress first developed a principled approach in bioethics (*principlism*). Beauchamp and Childress explicitly believed that utilitarianism and deontology were inadequate frameworks to effectively deal with the complexities that arise in medical decision-making. Their book *Principles of Biomedical Ethics* (2013) proposes four *prima facie* principles that, to this day, make up the core of much of bioethical reasoning and theorizing. Let us briefly touch upon each of these:

1. *Non-maleficence* (consequentialist): You should not cause harm. Application: Do not give children medicine that might be effective for their complaints but will cause more significant problems.

2. *Beneficence* (consequentialist): You should do good. Application: Try to cure patients.

3. *Autonomy* (deontological): Respect for people's autonomy means respecting their choices and enabling them to make informed choices by providing objective and complete information.

4. *Justice*: Treat patients fairly. Application: Doctors should not discriminate based on gender or race.

As these are *prima facie* principles, they often conflict and should be weighed against one another. For example, chemotherapy causes damage to the human body. If we strictly follow these duties as if they were obligations, the principle of non-maleficence would not permit the use of this type of treatment. However, most would agree that it is more important to try to cure someone. In this case, then, the principle of beneficence overrides the principle of non-maleficence. We might also encounter tensions between beneficence and autonomy. Consider the following: if a patient is unconscious and needs to undergo a surgical procedure to save their life, the physician cannot respect the patient's autonomy. However, in this case, the doctor may perform the procedure without consent; the duty to save a life is more important than respecting the patient's autonomy.

Principlism: comments and concerns

Before we move onto more specific applications of medical and health care ethics, it is worth reflecting on the presuppositions of principlism. Although 'the principles' to this day can be considered the primary 'toolkit' in much bioethical literature, they have received ample criticism within other disciplines and traditions. Two important criticisms have been widely expressed. First, postcolonial and feminist authors have expressed concerns over the presentation of the principles as the universal basis for moral reasoning and decision-making. Indeed, Beauchamp and Childress suggest that their principles accord to a *common morality*. They argue against the relativism gaining ground in ethical theorizing, suggesting that the principles are actually based upon something we all share—namely, a "set of norms shared by all persons committed to morality" (2013, p. 3).

Postcolonial and feminist thinkers have suggested that the principles presented by Beauchamp and Childress are deeply *situated*. These concerns reflect an argument against the representation of principlism as a rational and universal core of morality. Instead, these authors argue that the principles are committed to a distinctly *Western* view of what is of value. In particular, the emphasis on (individual) autonomy implies a commitment to specific views on moral agents and the role of communities. In some cultures, decision-making is not merely a personal endeavour but may be collaborative or even delegated to an elder. Researchers conducting clinical trials in such cultures may face the problem that the ethical procedures imposed by Western ethics committees— requiring each participant to sign their own informed consent—may be foreign to the local population. *Decolonizing ethics* presents a meaningful alternative to principlism. It entails developing and deploying more indigenous approaches to professional ethics rather than approaching local issues through a Western lens. In her article *Decolonizing Ethics* (2018), Amohia Boulton describes her work in a tribally-owned health research centre in New Zealand. The researchers at that centre use research principles from Māori protocols rather than Western protocols. She writes that there are Māori values or ethics that all Māori understand throughout the country. These include intrinsic or implicit principles such as *Whanaungatanga* (kinship or relationship), *Ahwi* (to cherish), and *Kotahitanga* (solidarity), which guide how people work together as Māori and how they treat one another. There are also explicit principles that are written down in strategic documents. These include *Rangatiratanga* (self-determination), *Manaaki Tangata* (care of all people), *Hauora Tangata* (health of the people, interpreted holistically), *Mātauranga* (knowledge), and *Ngākau Tapatahi me te Aurere* (working with integrity leads to achievement of purposes).

A second oft-cited worry is that the principles might be too abstract to offer actual guidance in decision-making; they do not describe *how* to act. Beauchamp and Childress (2013) defend the abstract nature of the principles by arguing that their lack of concrete content is precisely what allows us to apply and specify them according to the details of the relevant situation. Nevertheless, as we have seen in the context of care ethics, ethical concerns often involve concrete needs and relations with specific others. The 'objective' approach of the principles leaves little room for questions on personhood and experience.

Proponents of narrative ethics offer one potential alternative or supplement to principlism. Narrative ethics emphasizes the importance of storytelling, voice, and the first-person perspective. Authors such as Rita Charon (2002) and Arthur Frank (2013) recognize that human beings often understand and communicate moral experiences through storytelling. In healthcare, narratives can include the stories of patients, families, and healthcare providers. Such stories may offer richly detailed, personal, and contextualized accounts of the situation at hand and, as such, can provide valuable insights into the experiences of and relations between those involved. In addition, this approach is sensitive to the importance of listening. This is important, since patients often do not 'feel heard' by their caregivers. In sum, narrative approaches generally highlight the fact that, in ethical reasoning, we are dealing with questions of personhood, culture, and (personal) history—domains that are significantly underrepresented in the abstract and universalist approach of principlism.

Concerns have been expressed over narrative ethics as well. Some wonder whether stories are too subjective to feature in ethical (and clinical) reasoning. Others suggest that patients may not always have a clear idea of what they value, and that outside forces and narratives can influence personal stories. Some worry that given the high-speed, high-stress environment in which medical decision-making occurs, allowing stories 'to breathe' might be overly time-consuming.

Other critics offer a more measured response and suggest that a principlist approach can be fruitful to ethical reasoning, but indicate that the list provided by Beauchamp and Childress is too limited. The principles of autonomy, beneficence, non-maleficence, and justice may not be sufficient to deal with all ethical problems that might occur. They should be supplemented by dignity, integrity, vulnerability, and solidarity as essential principles in bioethical decision-making. Despite these worries, principlism with its emphasis on autonomy is still a central framework which a lot of ethical reasoning is based on, within the clinical context and biomedical research.

Now that we have a clear view of the ethical toolkit at the disposal of the medical or clinical ethicist, we can delve into some important topics within health care ethics.

Medical and clinical ethics: the patient-physician relationship

Starting with a range of questions we can categorize under the rubric of 'clinical ethics', let's take a closer look at some moral difficulties arising in the patient-physician relationship. This relationship is morally significant, not least due to its asymmetrical nature. Patients often present themselves to physicians in a state of (physical or mental) vulnerability. The physician, in turn, is situated as the expert, conditional for the receival of proper care or treatment. As such, they occupy the powerful position of effectively standing between the patient and their access to appropriate care. This imbalance evidently raises some fundamental ethical questions.

Paternalism and informed consent

Not that long ago, the relationship between doctor and patient could be characterized as *paternalistic*: patients (mostly) had to follow *doctor's orders*. We generally do not accept these overt forms of medical paternalism anymore. Current procedures and legislation in contemporary medical practice encode an important role for patients. Today, it is generally deemed unacceptable for physicians to act on behalf of their patients. One of the most important tools against medical paternalism is the requirement for *informed consent*.

The practice of informed consent stipulates that before patients can be admitted to medical procedures, they must agree (verbally or in writing) to the proposed treatment plan. As such, informed consent effectively ensures that patients can engage in autonomous decision-making without the coercion or influence of healthcare professionals.

Merely having the choice to accept or refuse treatment does not suffice to speak of proper consent. To enable genuine self-determination concerning treatment decisions, patients must have (or be given) access to all relevant information, including the benefits, the (physical, psychological, and potentially financial) risks of treatment, and potential alternatives to the proposed therapeutic action. In addition to respecting the patient's decision, physicians and other healthcare professionals should communicate openly and transparently with the patient so as not to impeach their autonomy and, relatedly, so as not to counteract the principles of beneficence and non-maleficence. Additionally, Onora O'Neill (2002) suggests that in order for informed consent to involve truly autonomous decision-making, patients should have access to meaningful alternatives. Sadly, this is often not the case, which raises questions on whether informed consent truly enables autonomy or merely acts as a legal tool to waive the responsibility of clinicians and hold patients accountable instead.

Substituted judgement

Whatever your stance on the ethics of informed consent, sometimes a particular medical situation renders it impossible to ask for direct consent. For example, children, unconscious patients, or those with severely diminished mental capacities may not be in a position to sign the relevant forms, or they might not be capable of understanding all the relevant information. Where does that leave us with regard to their health care?

Individuals who are not competent to consent are, of course, also eligible for medical care. In those cases, an appointed guardian should consent by proxy. This decision can be based on a *substituted judgement standard* when values or interests are known—for instance, when someone is in a coma, but their spouse knows what they would have wanted. If this is not possible, a proxy should consent with the best interest of the patient or research participant in mind.

Nevertheless, every effort should be made to inform the patient anyway. For example, those who are legally unable to consent should be asked for informed *assent*. Children should be asked for their opinions about research participation or treatment, which should be considered and featured in the final decision.

The patient's best interest

Other complications arise when physicians believe that patients act in opposition to what is in their best interest. A physician might think a patient should not undergo a risky operation that will only have marginal or even adverse effects on their quality of life, yet the patient is willing to take the risk.

Some authors have suggested that health care professionals can *nudge* their patients. Nudging refers to practices intent on influencing the patient's decision. For example, the 'framing effect' is a well-known nudging technique. Social psychology suggests that, for example, communicating that 25% of patients experience complications, rather than saying 75% do not, might push the patient's decision towards refusing treatment. While nudging can help patients reconsider what is in their interests, it is generally agreed that it conflicts with autonomy. Whether this constitutes an unethical transgression of informed consent is up for debate. Some ethicists, taking a utilitarian approach, think nudging can be morally permissible if the benefits clearly outweigh the risks. Others argue that nudging involves misleading the patient and is thus not permissible under informed consent requirements.

Another complication arises when physicians *perceive* that patients are making decisions based on 'irrational beliefs'. An example often repeated in bioethics courses is the refusal of a life-saving blood transfusion based on religious beliefs by a patient raised as a Jehovah's Witness. Some ethicists suggest that irrational beliefs—which might include religious beliefs, in their view—can affect a patient's capability to consent, and physicians should not merely respect the patient's judgement in such cases. Instead, physicians have a moral obligation to engage the patient rationally and discuss all the medically relevant aspects of the decision. In short, in addition to respecting patient autonomy, physicians should not be mere passive compliers to patient decisions. Instead, they are morally obligated to act as normative guides to help patients make the right decision.

As feminist scholars have emphasized, what makes a particular belief 'irrational' is highly contingent and based on contextual, social, and cultural factors. Thus, it might be that what seems irrational to a physician is of genuine importance to a patient. Therefore, in debates on autonomy and clinical decision-making, we should exercise caution in quick attributions of incompetence or irrationality. Nevertheless, conceiving informed consent not merely as a contractual obligation but as an opportunity for dialogue and deliberation aimed at mutual understanding seems to be a fruitful approach to the aforementioned concerns.

Liver transplant

You are an ethicist at a major hospital in Brussels. The transplant team has requested an ethics consultation regarding Marco, a thirty-nine-year-old Italian man with acute liver failure. Marco had spent several years working in South America before recently relocating to Belgium to stay with his cousin. He has no health insurance, no official residency in Belgium or Italy, and no current source of income. Marco's medical history includes substance abuse, though he states he has been sober for the past nine months. His condition is critical, and he needs a transplant within forty-eight hours.

> A matching liver has become available, but the procedure and follow-up care are expected to cost over €150,000, a sum the hospital would likely have to absorb. The administration is hesitant, as this expense represents the entire annual charity care budget, which typically supports dozens of uninsured patients. Concerns have also been raised about Marco's ability to adhere to the strict post-transplant regimen without insurance or stable living arrangements. Marco's cousin has committed to providing him with a place to live and helping with his recovery, but cannot contribute financially. She passionately advocates for him, citing his sobriety and determination to rebuild his life as reasons why he deserves this opportunity.
> What would the best course of action be in this case?

Ethics of medical AI

An important topic in contemporary medical ethics pertains to the use of artificial intelligence (AI) algorithms across a variety of medical applications. AI is increasingly used to optimize health expenditure and resource allocation, in diagnosis and risk prediction, and in patient and hospital management. Unsurprisingly the introduction of technically complex, highly-performant algorithms in sensitive contexts such as healthcare, biomedical research, and public health raises important ethical questions. This section surveys three central topics in contemporary AI ethics: ethical principles and regulations, algorithmic bias, and the role of AI in clinical decision-making.

AI ethics and governance

At present, various governing bodies are setting up systems for the governance of AI. How this is approached differs across the world. For this coursebook, we limit our discussion of regulations to the European context. The EU's approach to AI governance is to promote the uptake of 'human-centric and trustworthy' AI systems, which serve humanity and the common good (human-centric), and are lawful, ethical, and robust (trustworthy).

The two most relevant EU policy documents on AI are the non-binding *2019 Ethics guidelines for trustworthy AI of the High-Level Expert Group on Artificial Intelligence* (AI HLEG), and the binding *AI Act* (Regulation (EU) 2024/1689). We discuss these in order.

Ethics guidelines for trustworthy AI of the High-Level Expert Group on Artificial Intelligence

The AI HLEG guidelines set out a framework for trustworthy AI, stipulating seven requirements rooted in ethical principles, and European fundamental rights.

AI HLEG lists four ethical principles: respect for human autonomy, prevention of harm, fairness, and explicability. To make the guidance more concrete, the expert group has translated the ethical principles into seven requirements to be continuously evaluated and addressed throughout the development, deployment, and use of trustworthy AI. We will briefly discuss these principles and their related requirements in more detail. For complete descriptions of the principles and requirements, we refer to the AI HLEG guidelines document.

Respect for human autonomy: Humans interacting with AI systems should be able to retain self-determination, and AI systems' work processes should be subject to human oversight. This principle translates to the requirement of *human agency and oversight*. Users should be given sufficient information on the AI system to make autonomous decisions. Moreover, they have the right to involve a human in the decision-making process (i.e. 'human-in-the-loop') if they would be significantly affected.

Prevention of harm: AI systems cannot (exacerbate) harm to the dignity and integrity (mental and physical) of human and non-human beings, and the natural environment. Specific attention must be paid to vulnerable people and contexts of asymmetry in power or information. This principle requires developers to strive for *technical robustness and safety, privacy and data governance* and *societal and environmental wellbeing*. AI systems should be reliable, accurate, secure, and resilient to attacks, and they should have fallback plans in place and guarantee privacy or data protection. Data collected about individuals cannot be used to unlawfully or unfairly discriminate against them.

Fairness: The development, deployment, and use of AI systems should be fair. This means ensuring equal and just distributions of benefits and costs, and preventing unfair bias, discrimination, and stigmatization against individuals or groups. Individuals should be able to effectively contest AI decisions, and redress mechanisms should be in place in case of harm. The importance of fairness is stressed by the various requirements implicated, including *privacy and data governance, diversity, non-discrimination and fairness, societal and environmental wellbeing* and *accountability*. All affected stakeholders (humans, non-humans, society, the environment) should be considered and involved throughout the entire AI system's process. Clear and transparent oversight procedures should prevent unfair bias in datasets and development. AI systems should also be user-centric, accessible, and have equitable outcomes for persons regardless of age, gender, ability, or other characteristics. Finally, mechanisms should be in place to ensure responsibility and accountability for AI systems and outcomes before and after their development, deployment, and use.

Explicability: Explicability encompasses the need to have transparent AI processes, open communication on the systems' capabilities, purposes, and specific aspects, and the ability to explain the AI process and its decisions to those affected (explainability). This principle translates to *transparency* requirements for all elements relevant to AI systems. Datasets, technical processes, and outputs should be traceable and explainable, including related human decisions. Humans should be informed when interacting with an AI system and, where necessary, have the option for human interaction instead.

These principles are reminiscent of the bioethical principles of autonomy, non-maleficence (*harm prevention*), and justice (*fairness*), with the addition of the AI-specific principle of *explicability*. They are also subject to similar critiques. Tensions and conflicts may arise between these principles, sometimes in such a way that an acceptable balance cannot be achieved. For example, sometimes explicability can affect an algorithm's performance, which may lead to preventable harm. As such, these ethical principles have to be interpreted and translated into workable tools and applications.

AI Act (Regulation (EU) 2024/1689)

While the AI HLEG Ethical guidelines are helpful for AI systems' governance, they are legally non-binding. Regulation (EU) 2024/1689 (AI Act) lays down a risk-based legal framework for AI governance. In contrast to one-size-fits-all governance frameworks, risk-based frameworks aim to counteract overregulation by setting requirements and obligations proportional to the risk. In this way, the EU strives to find an optimal balance between (the benefits of) AI-related innovation and protection of health, safety, and fundamental rights against harmful effects. The approach may also be more flexible to govern quickly changing AI technologies, as it links obligations to the potential harms and risks, not specific technical classifications.

The AI Act outlines four risk levels. First, systems may be of *unacceptable risk* when "considered a clear threat to safety, livelihoods and rights of people" (European Commission, n.d.). Categorization systems using biometric information for the inference of sensitive and protected characteristics—such as race, political opinions, and sex life—fall under this category. These systems are banned under current regulation. *High risk* systems have the potential to cause significant harm to the health, safety, or fundamental rights of individuals or the environment if they fail or are misused. These systems are strictly regulated. *Limited risk* systems have specific transparency obligations due to, for instance, the risk of manipulation or deceit. Developers and deployers must ensure that users are aware that they are interacting with AI. Generative AI applications (e.g. GPT, CoPilot, and Claude) generally fall into this category. All other AI systems (e.g. spam filters) are considered *minimal risk* and are not subject to any mandatory legal requirements or obligations.

Governance of medical AI

The AI HLEG ethical guidelines and the AI act are directed at all AI systems, not specifically medical AI. Consider an AI model implemented in a (non-invasive) wearable health tracking device (e.g. a smartwatch), which continuously monitors vital signs such as blood pressure, skin temperature, and heart and respiration rate. These vital signs are input for an AI model which would alert the

> patient and a trusted contact person (family member, nurse, etc.) when medical attention is needed.
>
> How do the AI HLEG ethical principles and requirements apply to this AI system? Which aspects of this system could be ethically problematic? How would you address it? In which risk category of the AI Act would you place this AI system?

Algorithmic bias and justice

As we saw earlier in this chapter, one of the central principles in biomedical ethics is justice. Physicians, other healthcare professionals, and institutions ought to treat patients fairly and refrain from discrimination based on social categories such as gender, race, religion, or socio-economic status. As some recent controversies have shown, medical AI systems deployed across various contexts of healthcare put the matter of justice and *fairness* at the forefront of current ethical discussions.

In 2019, Obermeyer and colleagues published a study in *Science* exposing significant racial bias in a widely used healthcare algorithm for identifying 'high-risk patients' for additional clinical management. The algorithm disproportionately excluded Black patients from needed care, even when they had more severe health conditions than their White counterparts. Another example is the GAIL Risk Score algorithm used in breast cancer risk assessment. Although widely used, its performance on younger populations, non-Western patients, and atypical breast cancers is increasingly shown to be subpar. These issues point toward persistent biases in medical algorithms. *Algorithmic bias*—or systemic distortions or unfair influences in AI decision-making processes disproportionately favouring or disadvantaging particular individuals or groups—is a particularly pressing issue for healthcare.

The roots of algorithmic bias often lie in the data used to train and test AI systems. A common mantra in computer science—'garbage in, garbage out'—captures this problem succinctly: if your data is of low quality, expect low quality results. In the context of healthcare, the problem of missing and low-quality data evokes a larger history of ethical misconduct. Historical underrepresentation and exploitation of marginalized groups—such as women, people of colour, and disabled people—in biomedical research has resulted in datasets poorly reflecting the diversity of real-world populations. These and similar dynamics have resulted in a lack of reliable data today, which impacts the performance of medical algorithms. However, bias is not merely a matter of *missing* data. Most health data is, and historically has been, generated in the context of routine healthcare. As is well-documented, healthcare professionals often hold implicit biases towards marginalized patients. These biases, which can lead physicians to dismiss the concerns of certain groups, are embedded into datasets and perpetuated by AI systems.

'Garbage-in, garbage-out' tells only part of the story of bias in AI, however. In their book *Data Feminism*, Catherine D'Ignazio and Lauren F. Klein (2020) show that data are never truly 'raw' but rather are shaped and moulded by the social and political context in which they are generated. As philosopher Gabbrielle Johnson (2021) has shown, existing (unjust) social and political structures infiltrate algorithmic systems in surprising ways. Recall the resource allocation algorithm discussed by Obermeyer and colleagues. The algorithm did not directly use the patients' race as a feature in determining health needs. Instead, their analysis revealed that healthcare spending was taken as a measure for health needs. On the face of it, it makes sense to conclude that those patients spending more on healthcare, generally, have higher care needs. However, because race is a predictor of improper access to care in the US healthcare system, the algorithm implicitly incorporated these racial biases. Mechanisms like these demonstrate how societal biases seep into AI systems, even when sensitive categories like race or gender are excluded.

Clinical decision-support systems and the patient

While algorithms can be used for resource allocation, the applications we will likely face most directly are clinical decision-support systems (CDSS). These AI-driven tools aid physicians in diagnosis, risk prediction, and in making treatment decisions. Though CDSS offer potential for improving healthcare and can process vast amounts of health data, identifying patterns beyond individual practitioners' capabilities, their introduction into the clinical encounter raises several ethical questions as well.

The clinical encounter typically provides a space where patients actively engage with healthcare professionals, seeking explanations for treatment recommendations and advice based on their specific circumstances. This highlights the importance of *transparency* and *explainability* for medical AI – healthcare professionals must be able to understand and explain AI-generated recommendations to patients, particularly when they disagree with the system's conclusions. This transparency is also essential for maintaining patient *autonomy* and ensuring informed consent. In this context, the implementation of CDSS should be viewed through and designed according to a *collaborative* lens (e.g. human-in-the-loop), where such systems provide additional, explainable input to support clinical decision-making, but are not relied on exclusively.

Another concern involves algorithms potentially dominating patient-physician dialogue. AI systems' perceived 'objectivity' may diminish the patient's voice in clinical decision-making. Philosopher Miranda Fricker (2007) uses the term *epistemic injustice* to describe situations in which a patient's knowledge and testimony about their own condition(s) is dismissed. Such dismissal is ethically problematic for two reasons: it undermines human dignity by denying the patient's role in knowledge creation about their own health, and it can lead to practical harm when important patient concerns are overlooked in treatment decisions. We risk amplifying these existing concerns if we allow CDSS recommendations to overshadow patient experiences. AI systems can only capture particular aspects of disease, primarily those that are easily quantifiable. As such, while an AI system is well-suited to incorporate genomic data or the results from blood tests, it will struggle in capturing the social and experiential realities of illness. Effective clinical decision-making must consider not only pathophysiological factors but also patients' values, social relationships, and lived experiences with illness and treatment. Failing to incorporate these qualitative aspects while deferring to AI recommendations would represent a new form of *medical paternalism*. As we saw throughout this chapter, AI ethics borrows heavily from principles in biomedical ethics. While these are important tools for AI assessment, more fundamental questions may need to be addressed as well. As feminist philosopher Alison Adam reminds us in her book *Artificial Knowing (1998)*, the implementation of AI is preceded by important social and cultural questions: what role *can* and do we *want* AI to play in our structures and institutions? Adam – while critical – points toward the potential for AI systems to provide care, alleviate critical workers from cumbersome tasks, and democratize our access to knowledge. AI is not necessarily a threat to existence, nor a solution to all our problems, but first and foremost, it is a useful tool that can mean something *for all of us*.

Reproductive ethics

Introduction to reproductive ethics

In reproductive ethics, people have expanded upon Beauchamp and Childress' principles and applied them to ethical questions in conception, childbearing, and rearing.

On a more fundamental level, it is essential to note that opinions on the status of the unborn child heavily influence this debate. When does an embryo or foetus become a person? For some, this happens right after conception, as everything is available for the embryo to become a person. For others, this is when the embryo has implanted, the nervous system has started developing, or when the foetus is viable after twenty-four weeks. The law on abortion in The Netherlands, for example, takes the foetus' viability as a starting point and forbids abortion after twenty-four weeks.

For some philosophers, a human is only a person after birth. It goes without saying that how one perceives the 'person status' of unborn life heavily influences what one believes can morally be done to it. For example, people who believe an embryo is a person or a potential person right after conception may object to embryos being thrown away during fertility treatment. In the case of embryo selection, which is offered to prospective parents who know they carry a genetic disease, several embryos are created and checked for genetic mutations. Only an embryo without the mutation is transferred back to the prospective mother's womb. The rest are not used for fertility purposes. Some have argued that in this case, gene editing an embryo to correct the mutated gene (with technologies such as CRISPR/CAS9) might be better, as it would not necessarily involve creating embryos which will not be used. So, this would be a solution for people who object to creating more embryos than needed. Of course, this does not change the fact that many non-viable embryos were used or even explicitly created to be tested on and then destroyed in order to develop this technique.

Besides debates on moral agency, reproductive ethics also focuses on particular biomedical technologies related to (human) reproduction. As we will see in chapter 6, developments in epigenetics raise new questions on reproductive and parental autonomy. Embryo selection is another recent technological development that requires us to reassess longstanding debates in reproductive ethics. Embryo selection is a technique to select embryos without genetic defects by conducting a genetic test on the *in vitro* embryo (official name: Preimplantation Genetic Testing). The technique has advanced to make whole-genome sequencing of the embryo's genome possible. Bioethicists have considered the extent to which such embryo selection should be allowed. Maybe it should be allowed to prevent serious harm, such as specific congenital genetic diseases. Perhaps it should also be allowed to select embryos likely to develop diseases later in life, such as Alzheimer's.

Julian Savulescu (born 1963), an Oxford ethicist, has suggested the principle of *procreative beneficence* (2001). He has fine-tuned this principle in various articles. However, it boils down to this: prospective parents should, in principle, and if possible, choose the embryo that will develop into the child that potentially has the best life. He gives the example of IQ: if it is possible to select, *in vitro*, an embryo with an IQ of 140, one should do that. Many would intuitively feel that it may be better, or even a duty, to select the embryo that would not develop severe diseases if you have a choice. However, for characteristics such as IQ, this is much less intuitive. Can we decide how well someone will experience life based on their genotype? Does someone with a higher IQ necessarily have a better life? For whom is this better, for the person themself or their society? Many authors have criticized the principle of procreative beneficence, and the challenges associated with a utilitarian approach are also applicable here.

John Robertson (1943–2017) has taken a deontological approach to reproductive ethics, with the principle of procreative liberty (or reproductive autonomy) (1983). This principle states that anyone has the right to reproduce or not to reproduce. This means that people can choose whether they want to have children. For Robertson, this also means people may decide on the children they want. If prospective parents find it important that their children are musical, they may choose an embryo with a genetic propensity for perfect pitch. Not anything goes, however, as there are some limitations. The choices must not harm the resulting child. Robertson would have objected to parents choosing a child with a disease. This approach has its drawbacks. How can it be determined whether the parents' choice would harm the resulting child? What counts as a disease? In a gendered world, could the gender of a child be seen as a good reason not to want that particular child? Is it fair to children that they are picked based on the parents' preferences? Should parents not accept children as they are?

Reproductive rights and justice are also a central concern in feminist and disability scholarship and activism. Calling for reproductive justice, Black and Indigenous feminist thinkers have highlighted how, apart from the right to not have a child (which is put forward in the abortion debate), other rights are equally important: the right to raise your own child, and the right to do so in a safe and healthy environment. Silliman and colleagues (2004), for example, refer to the reproductive violence bestowed upon enslaved and Indigenous people, and the many ways these histories still play out today. Other reproductive justice scholars, such as Sigrid Vertommen and colleagues (2022), emphasize the social, cultural, and political assumptions underlying assisted reproductive technologies, and the individuals—often women—whose bodies and reproductive tissues are essential in medically assisted reproduction but who are nevertheless often neglected (surrogates, egg and sperm donors, etc.). In ethical debates on reproductive technologies, it is vital to be aware of who is being excluded from the discussion, but also who is encouraged or discouraged to procreate. One example is found in the research of Virginie Rozée and colleagues (2024), who analysed European fertility clinic websites and concluded that medically assisted reproduction is presented as primarily a matter for white, cisgender, and heterosexual women.

Towards a disability bioethics

A prominent debate concerns the ethics of 'choosing disability' in the context of preimplantation genetic testing and prenatal diagnosis in general. Some disability scholars suggest that the selection of embryos without disabilities is eugenic—i.e. aimed at 'improving' the population through genetic selection. Feminist bioethicists have lamented the decontextualization of the debate in much ethical theorizing (see for example Scully 2023). Arguing against simplified views of reproductive autonomy, they suggest that we should scrutinize to what extent pregnant people are genuinely free to choose, given the overall discrimination of particular social identities and the lack of social support. Another example of the tension between a feminist stance in favour of choice and sensitivity to disability justice is the letter exchange between Eva Kittay (born 1946) and her son Leo. Kittay (2019) emphasizes the autonomy of women while at the same time agreeing with her son that in our current society, some reasons for not choosing a future with a disabled child are more informed or better than others.

While most of these debates primarily concern the 'acceptance' of disability, the 'active preference' for a disabled child is also discussed. An oft-cited example involves a Deaf couple actively seeking out a sperm donor with the right kind of genetic deafness. While many were quick to condemn such behaviour, their decision might also lead us to reconsider our intuitions on disability. As disability scholar and bioethicist Jackie Leach Scully (born 1961) notes, these debates are ultimately premised on whether we consider (the choice for) disability harmful and, if so, whether it is severe enough to outweigh procreative autonomy (2008, 2022). As demonstrated by the responses to the aforementioned case, many clearly take this stance. We might, however, want to consider how, for many disabled people, 'disability' is deemed to be *merely* a difference, instead of necessarily a *bad* difference, to paraphrase philosopher Elisabeth Barnes (2018).

These applications of the principles in the context of reproductive decisions and technologies show that even if moral theories and principles are fine-tuned to tackle specific questions, merely choosing a principle and applying it is not enough: cases require weighing up principles and sensitivity to the particular circumstances in which questions arise. Notably, the concerns of the disability community also serve as a stark reminder of how intuitions may vary wildly across social groups. In addition to being contextually sensitive, our ethical theorizing should also involve the input of the relevant stakeholders. Maybe one of the primary tasks of the bioethicist is to make sure that those voices are heard that have traditionally been left out of the debate, but that are often most affected.

Public health ethics

As we have pointed out earlier, questions on public health also fall within the purview of health care ethics. Public health interventions such as public sanitation, fluoridization of community water, and vaccine policies have historically been some of the most effective ways of improving or maintaining population health. Two more recent examples are the lockdown measures and the vaccination campaigns in the early stages of the COVID-19 pandemic. Public health initiatives differ from traditional, clinical health care since they primarily aim to advance health at the *level of the population* through enacting top-down policies and measures affecting individual citizens. It is evident that this might raise some important ethical questions.

Triage and resource allocation

Vaccine distribution

A global pandemic is happening due to a newly mutated virus. Infection can be lethal regardless of age, and all sorts of patients are being admitted to emergency care facilities, putting enormous pressure on the healthcare system. Countries decide on different policies: some have ordered lockdowns, while others expect that people will take appropriate measures themselves to ensure everyone's safety. In some countries, people keep living as if nothing had changed; while in other countries, people take precautions.

After a few months, scientists have succeeded in developing an effective and safe vaccine, the only available treatment for the infection. However,

> given its complexity, the vaccine is scarce, and only limited amounts can be produced in highly specialized facilities. The distribution of the vaccine becomes a critical task for global health, and the World Health Organization (WHO) sets up a working group to decide on the allocation of the vaccines for each country. As an expert on the matters of vaccines, infectious diseases, and public health, you are, of course, invited on this committee. What would be a just distribution? Which criteria do you use to justify your choice? Which factors do you take into consideration?

Health care and research funding are not unlimited as they compete with resources for other social goods. As a result of such competing interests, policymakers, researchers, and clinicians often have to focus on one issue over another. More extreme conditions of scarcity put these issues at the centre of ethical decision-making. A recent example was the allocation of hospital beds and respirators in the early days of the Covid-19 pandemic.

During the pandemic, treatments, hospital beds, and even health care professionals were in short supply. This resulted in policymakers and physicians having to make difficult decisions on who gets a bed at the intensive care unit, who receives a respirator, and who doesn't. Evidently, this may lead to significant moral distress in health care workers. In response, many countries developed so-called *triage protocols* to guide physicians in decision-making.

The general idea of 'triage' is utilitarian. Given the limited resources we have at our disposal, how and according to which criteria can we maximize the lives saved? What seems like simple calculus raises some important questions about which principles to follow in triage. Most guidelines take the 'requirement of critical care' as uncontroversial. Instead of merely following a 'first come, first serve' approach, we generally would want to only offer a bed to those patients who genuinely need it. Discussions among ethicists concern primarily which *other* criteria to rely on to prioritize certain patients over others.

Age is often taken as one of those criteria. While relatively uncontroversial, some ethicists have suggested the *ageism*—discrimination based on age—presupposed here can be discriminatory and, therefore, unjustifiable. More controversially, some have suggested that doctors and politicians, for example, should receive preferential care since they might significantly impact general well-being. Again, others suggest that lifestyle should be featured in the decision-making process. Should a lifelong smoker, for example, receive a respirator? Most of these latter criteria are (rightfully) controversial.

What seems perhaps less contentious, and more objective is relying on the likelihood of survival post-hospitalization. In addition to epistemological questions such as 'how do we assess this likelihood?' and 'where do we set the threshold?', this raises some clear moral quandaries. Tolchin, Hull, and Kraschel (2021) note that while comorbidities (other conditions such as obesity, diabetes, or hypertension that affect the clinical outcome of COVID-19 infection) seem reasonable and objective parameters to base triage on, comorbidities often also track social determinants of health. Suppose we take obesity as a condition leading to a lower rank on the respirator list, and obesity is correlated to lower socioeconomic status. In that case, our 'objective' protocol might disproportionately affect marginalized groups, raising justice-related concerns.

One way to express post-hospitalization survival is through quality-adjusted life years or QALYs. QALYs represent the number of years lived in 'full health'—i.e. without disability. The general idea is that while some treatments might prolong life, they might also lead to disvalued states of well-being. So, instead of just measuring the effect of a particular intervention on life expectancy, these metrics are calibrated in terms of quality of life. Quality is a notably subjective term, of course. To arrive at a widely supported notion of quality, QALY's are thus generally measured by surveying the general public on how they would value particular, hypothetical health states. Like other forms of utilitarian calculus, policymakers rely on these evaluations to maximize the calculated number of QALYs gained through a specific intervention. As philosopher Laura Cupples (2020) rightfully points out, the idea of QALYs is built on the *ableist* assumption that rational people would prefer a shorter life in an able-bodied state than a longer one lived with disability. Cupples suggests that this is further corroborated by how QALYs are measured. Given their situatedness as (generally) able-bodied, the general public might not have a nuanced view of life with disability. Instead, Cupples argues—in line with feminist epistemologists—that we should primarily ask disabled people to evaluate these states since their testimony from experience might be more objective than that of the general public.

We might also give up on looking for a utilitarian calculus altogether. Strict egalitarians, for example, argue that no differences between patients can be operationalized as reasonable criteria for differential care allocation. Instead, they favour randomized processes – as in clinical trials, as we will see later – since these exclude external factors from triage. Finally, prioritarians intentionally favour those patients who are worse off. Instead of pursuing likely benefits, as a utilitarian might, prioritarians position sickness or socio-economical disadvantage as relevant features in prioritizing care.

Prevention and health promotion

Another increasingly important topic in public health ethics is prevention. Many commentators suggest that one of the central challenges of contemporary healthcare is the increasing costs due to an aging population and a steep rise in chronic conditions. Unsurprisingly, these past few decades have been marked by a push towards a preventive model for healthcare.

Policy-makers support these efforts by enacting health promotion policies. Banning cigarettes from public spaces, promoting exercise and healthy diets through public campaigns, and national screening or vaccination programs are all examples of such top-down health promotion policies.

Many of these programs generally frame health promotion and prevention in terms of 'making the right choices'. Relatedly, we often find conditions like diabetes, obesity, and cardiovascular diseases referred to as 'lifestyle' diseases. It is suggested then that these conditions are (to large extents) avoidable. Health promotion campaigns often aim to change the behaviour of individuals so that they adopt healthier lifestyles.

As philosopher Per-Anders Tengland (2016) argues, while such interventions may lead to better outcomes from a public health standpoint, such campaigns are often paternalistic. Indeed, in encouraging citizens to make particular health-related choices and behavioural changes, professionals assume and impose specific understandings of the relevant problems and good health-related behaviour. One reason why health promotion campaigns often fail to attain the desired outcomes is that the (lifestyle) issues identified are considered less relevant or important to people. Instead, people care about their quality of life, which is not entirely reducible to their health.

Additionally, these campaigns might also foster stigmatization of particular behaviours or bodies. Being overweight may be considered risky or irrational behaviour which the individual is to be blamed for. This might, in turn, divert attention from social explanations such as limited access to healthy food or open space for exercise, which might equally affect one's opportunities to follow such directives.

Finally, conceiving of prevention in terms of behavioural change holds individual patients responsible for conditions which might be better tackled by addressing social causes. Food deserts—areas where healthy, affordable food is scarce—are often located in poorer areas. Instead of investing in promotional campaigns, communal development and improving access to nutritious, inexpensive meals might be a more effective way to improve public health.

Research ethics in biomedical research

Research on human subjects is imperative to gain insights into the pathophysiological mechanisms of disease, discover and validate new treatments, and monitor their effects on patients. It is also clear that these research practices may be subject to relevant ethical questions. For example, we have already seen that respecting patient autonomy entails informing them about the treatment or medical procedure and acquiring consent.

In most countries, researchers must submit prospective research and trial designs to an Institutional Review Board or Ethical Commission. Sometimes, researchers feel that these requirements are burdensome. They suggest that these ethical constraints hinder research and, thus, scientific progress. Throughout the rest of this chapter, we will examine some of these requirements and how good science and ethical science can, should, and generally do go hand in hand. But first, to clarify where these ethical requirements come from, we will give you some examples of where research went wrong.

Why research ethics matters

The Tuskegee syphilis experiment ran from 1932 to 1972 in Tuskegee, Alabama. It aimed to study the natural development of syphilis (Jones, 1993). The experiment enrolled 600 Black men, 399 with syphilis and 201 without syphilis. The men with syphilis were not told that they were part of an experiment or that they had syphilis. They were told they were treated for 'bad blood'. The participants did not receive any treatment. As part of their participation, they did receive free medical exams, meals, and burial insurance. In 1947, penicillin became the drug of choice to treat syphilis, but researchers still did not offer it to participants. In the experiment, 128 subjects died, 40 women contracted syphilis, and 19 children were born with congenital syphilis. In 1972, The Washington Post reported on the experiment, and in 1973, there was a class-action lawsuit. In 1974, there was a ten-million-dollar settlement, and the US government promised lifetime medical benefits and burial services to all living participants. In 1997, President Clinton apologized on behalf of the Nation, and in 2004, the last participant died.

The atrocities of the nazi experiments are well known. One example is the nazi freezing experiment. In 1941, German soldiers were confronted with cold weather on the Eastern Front (Annas, 1992). So, Ernst Holzlöhner and Sigmund Rascher wanted to know how much cold humans could tolerate. They performed 360–400 experiments on 280–300 Jews in Dachau and Auschwitz. Participants had to sit in cold water to see how long they would last and how they could be 'reheated'. Approximately 100 participants died during these experiments.

Clinical research ethics

While these examples are primarily historical, it is essential to note that even today, research participants—especially those who belong to marginalized groups—are vulnerable to exploitation for scientific (or financial) gain. In their text *What Makes Clinical Research Ethical* (2000), Emanuel, Wendler, and Grady (2000) summarize this intrinsic issue of biomedical research well: "By placing some people at risk of harm for the good of others, clinical research has the potential for exploitation of human subjects. Ethical requirements for clinical research aim to minimize the possibility of exploitation by ensuring that research subjects are not merely used but are treated with respect while they contribute to the social good" (p. 2701). Before we discuss the principles they laid down, let us try to find some of them ourselves.

> **Discussion:** If you were to design an ethical code of conduct for researchers conducting experiments with human persons, what should be included?

Emanuel, Wendler, and Grady devised seven requirements to aid in assessing ethical research:

- Value
- Scientific validity
- Fair subject selection
- Favourable risk-benefit ratio
- Independent review
- Informed consent
- Respect for enrolled subjects

The first requirement, *value*, states that a treatment, intervention, or theory should improve health and well-being or increase knowledge. To be evaluated as such is necessary for research to be ethical, considering the finite resources and risk of exploitation. In short, since funds are limited and could serve several other socially relevant goals, researchers should think hard about what their research is for. Determining what is of 'value' depends on our social aims and interests. While traditionally, clinical value was determined by those conducting research—primarily (male) scientists and clinicians—ethicists and policymakers are increasingly pushing for participatory research, where those most affected should have a say in identifying the research questions, aims, and outcomes.

Scientific validity requires that research be methodologically rigorous—meaning that accepted scientific principles and methods, including statistical techniques, should be used to produce reliable and valid data. While this might seem a clear-cut, scientific issue, ethical questions also feature here. For example, in clinical trials a p-value lower than 0.05 signifies statistical significance, but we may want to reflect on the consequences of accepting such an error rate. In some cases, falsely identifying 5% of patients as at risk of cancer might not be morally permissible; while a 5% error rate might be acceptable in an influenza test.

Additionally, clinical research should have *clinical equipoise*, meaning that research comparing therapies must have an honest null hypothesis: clinical researchers must genuinely not know which treatment is better. Also, placebos should not be used when conventional treatment is available. For example, since we already have good drugs for managing diabetes, new drugs need to show benefits compared to those established treatments.

The third requirement, *fair subject selection*, entails that scientific objectives (not vulnerability or privilege) and the potential distribution of risks and benefits should form the basis for selecting communities to study and the inclusion criteria for individual subjects. This means that the study must be generalizable to the populations that will use the intervention. It also means that those who bear the risks and burdens of research should be able to enjoy the benefits (*distributive justice*). This does not mean, however, that healthy controls should be excluded from all biomedical research. Improvement in the representation of children and women in clinical trials is still possible and necessary. Children are considered a vulnerable population. They are often underrepresented in clinical trials, although they still receive prescriptions for drugs that are not tested on children. Hence, they are *therapeutic orphans* because they are either denied the use of many new treatments or exposed to drugs that have bypassed rigorous regulatory evaluation. Women are also underrepresented in clinical trials, except for clinical trials investigating reproductive organs (this is termed *bikini medicine*, because it focuses on what is covered by a bikini). However, differences related to other aspects of the female body and endocrine system might also affect other drugs. Indeed, 80% of drugs withdrawn from the market are withdrawn because of side effects in women. For example, a dosing issue for women was only discovered after Ambien was on the market for twenty years, leading to early-morning car accidents in which women were predominantly involved.

A *favourable risk-benefit ratio* requires that risks must be minimized, potential benefits must be maximized, and benefits should outweigh the risks. This sounds straightforward, but balancing risks and benefits is complex and controversial. Can payment count as a benefit? How fair is it to balance societal benefits and burdens/risks to individuals? How do we define risk?

The *precautionary principle* states that if an action or policy has a suspected risk of causing harm to the public or the environment in the absence of scientific consensus (that the action or policy is not harmful), the burden of proof that it is not harmful falls on those taking the action that may or may not be a risk. Nevertheless, opinions on this principle are divided. To some, it is unscientific and an obstacle to progress. To others, it is an approach that protects human health and the environment.

The requirement of *independent review* entails that unaffiliated individuals review the research and approve, amend, or terminate it. This ensures that the potential impact of conflicts of interest is minimized, and social accountability is ensured. If you are conducting research, you must have this approved by an ethics committee. An ethics committee often comprises of physicians, specialists, nurses, ethicists, and philosophers. Some authors go further and suggest that institutional review boards must be sufficiently diverse. Relying on the argument from situatedness we saw earlier, they indicate that if moral intuitions and knowledge are related to who we are, a more diverse group is likely to better identify potential issues in research proposals.

This sixth requirement is *informed consent*. This means that research participants should be accurately informed about the research's purpose, methods, risks, benefits, and alternatives. They should understand this information and its bearing on their personal clinical situation (if applicable). They should consent to participate in the research voluntarily (without outside pressure) and, as noted earlier, be competent to consent. The randomization of assignment to treatment or a control group should be explained well to participants. Particular attention should go into averting the *therapeutic misconception. The therapeutic conception occurs when patients or research participants hold the* mistaken belief that their participation in a clinical trial will lead to personal benefit. This is clearly not always the case, for example, when patients are offered a placebo treatment, or the intervention turns out to be ineffective. Informed consent also means ensuring that participants know they are primarily participating in research to contribute to scientific knowledge rather than their own benefit.

Informed consent in biomedical research is sometimes presented as a blanket statement. The participant signs a document once at the beginning of the study. Afterward, they donate their samples or data to the researchers, which the research community can use forever without any restrictions. This is sometimes called the 'sign here to consent forever'-model. This model has become more contentious in the wake of the controversies surrounding the HeLa cell line (Skloot, 2010). The HeLa cell line was taken from African American woman Henrietta Lacks and distributed to research labs worldwide. The HeLa cell line is still used to this day. Neither Lacks (who eventually succumbed to cervical cancer) nor her family would receive compensation for the highly lucrative tissue sample taken from her.

In the current context, where health care research increasingly involves gathering large amounts of data, we must revisit such narrow interpretations of consent. As clinical data is increasingly used for secondary purposes (e.g. biobanks or reusing clinical datasets), this raises questions such as: to whom does this data belong? Should we consider data as donated? Or should participants have a say in what kind of secondary research their data are used for?

A recent example of an issue with informed consent concerns the *Havasupai tribe* in Arizona (Van Assche, Gutwirth, and Sterckx 2013). In 2003, the University of Arizona gathered blood samples from Havasupai members. The goal was to investigate the high incidence of type 2 diabetes—itself linked to historical food shortages due to forced relocation of the tribe by the US government—amongst the Havasupai people. The tribe members received oral information about the focus of the research project on diabetes, after which they willingly participated in the study and provided blood samples. In the written informed consent form, however, the purpose of the study was described more vaguely ("study the causes of behavioural/medical disorders"), so the research scope was not limited to diabetes only. One of the researchers involved had already obtained a research grant to study genetic causes of the (assumed) high incidence of schizophrenia within the Havasupai tribe. As a result, the tribe's genetic material, blood samples, and biomedical data were also used and shared with other researchers to research inbreeding and schizophrenia.

Additionally, the samples were used to trace the Havasupai genetic origin, contradicting their own cultural origin story, without seeking permission from the tribe. All these additional research aims were not adequately disclosed to Havasupai Tribe members. The Havasupai eventually sued the University of Arizona for invading personal and 'cultural and religious' privacy and causing harm and distress. Their blood samples were returned, and participants were financially compensated.

Examples like the Havasupai make us reconsider the type of consent we can demand from research participants. Some alternative models for informed consent are *tiered consent* (e.g. 'I consent to this research but not further studies') or *dynamic consent* (the participant has access to a digital platform to check what kind of research their samples are used for, and can revoke their consent accordingly). However, does *individual* informed consent suffice? Should community considerations not be brought into perspective?

The last principle is *respect for enrolled patients*. This means that a patient's privacy should be protected and they should be allowed to withdraw. It also means that their well-being should be monitored, and they should be informed about potentially relevant research outcomes for themselves and in general. The *European Union General Data Protection Regulation* (GDPR) came into effect to improve data privacy and protection in May 2018. The GDPR has the following principles: (1) consent for data usage and storage should be obtained; (2) if a breach of security and privacy has occurred, participants should be notified promptly; (3) participants have a right to access their data; and (4) participants have a right to be forgotten (to delete their data).

There is a right to *data portability*: the data subject has the right to receive personal data concerning them, and privacy should be part of the design. Also, researchers should designate potential data protection officers who are aware of the regulations.

Research in developing countries deserves special attention. We have seen that placebos should not be used in clinical trials if a known treatment is available. However, which standards of care should apply here, those of developed or developing countries? Some have argued that a placebo is justified in clinical trials in developing countries, even if the treatment is available in developed countries but not in developing countries. At the same time, we could argue that we owe more care to research participants in developing countries. In their 1997 publication, *"Unethical trials of interventions to reduce the perinatal transmission of HIV in developing countries"*, Lurie and Wolfe argue that certain clinical trials with Zidovudine in developing countries were unethical. Indeed, in 1994, there was the discovery of a significant reduction in HIV transmission from mother to child after treatment with Zidovudine (25% to 8%). However, this treatment was expensive (over $800 per pregnancy)—the WHO decided that a less costly alternative was needed for developing countries. A shorter treatment with Zidovudine was proposed. A double-blind, placebo-controlled trial with two arms was executed: one arm was a placebo, and the other was a shorter treatment with Zidovudine 076. The argument was that using the placebo arm was warranted here because the standard of care in the developing country was 'no treatment'. However, Lurie and Wolfe argued that this justification was invalid since the treatment was available in developed countries: foetuses were potentially exposed to HIV if they were in the placebo arm, which could have been prevented.

Conclusion

Although medical ethics is often reduced to a set of hot button issues such as end-of-life care and designer babies, this chapter showed that clinical encounters, public health, and biomedical research give rise to a variety of ethical questions. In order to address such a wide swath of complex situations, bioethicists generally rely on a set of prima-facie principles such as autonomy, non-maleficence, beneficence, and justice to assess their stakes and weigh the various values involved. While frameworks such as principlism often function as a useful starting point, throughout the chapter we have also seen examples of their limitations. Although still a foremost part of the bioethicist's toolbox, these principles themselves are situated in a specific historical, cultural, and social context. As such, they need to be enriched by considering the voices of different traditions and social positions. The need to consider existing inequalities and involve stakeholders from a variety of cultural or social backgrounds became even clearer when we discussed reproductive issues, public health, and algorithmic justice. Bioethicists play an important role in guiding policy-making and public discourse on these topics. As such, they have the opportunity or maybe even the obligation to ensure that all relevant voices are heard.

Bibliography

Adam, Alison, ed. 1998. *Artificial Knowing: Gender and the Thinking Machine*. London New York: Routledge.

Annas, George J., and Michael A. Grodin, eds. The Nazi Doctors and the Nuremberg Code: Human Rights in Human Experimentation. Oxford University Press, 1992.

Barnes, Elizabeth. 2018. *The Minority Body: A Theory of Disability*. Studies in Feminist Philosophy. Oxford: Oxford University Press.

Beauchamp, Tom L., and James F. Childress. 2013. *Principles of Biomedical Ethics*. 7th edition. New York: Oxford University Press.

Boulton, Amohia. 2018. "Decolonising Ethics: Considerations of Power, Politics and Privilege in Aotearoa/New Zealand". *Southern African Journal of Social Work and Social Development* 30 (1). https://doi.org/10.25159/2415-5829/3825

Camporesi, Silvia, and Maurizio Mori. 2021. "Ethicists, Doctors and Triage Decisions: Who Should Decide? And on What Basis?". *Journal of Medical Ethics* 47 (12): e18–e18. https://doi.org/10.1136/medethics-2020-106499

Charon, Rita, and Martha Montello, eds. 2002. *Stories Matter: The Role of Narrative in Medical Ethics*. Reflective Bioethics. New York: Routledge.

Cupples, Laura M. 2020. "Disability, Epistemic Harms, and the Quality-Adjusted Life Year". *IJFAB: International Journal of Feminist Approaches to Bioethics* 13 (1): 45–62. https://doi.org/10.3138/ijfab.13.1.03

D'Ignazio, Catherine, and Lauren Klein. 2020. *Data Feminism*. Cambridge: MIT Press.

Emanuel, Ezekiel J., David Wendler, and Christine Grady. 2000. "What Makes Clinical Research Ethical?". *JAMA* 283 (20): 2701. https://doi.org/10.1001/jama.283.20.2701

Engelen, Bart. 2019. "Ethical Criteria for Health-Promoting Nudges: A Case-by-Case Analysis". *American Journal of Bioethics* 19 (5): 48–59. https://doi.org/10.1080/15265161.2019.1588411

European Commission. n.d. "AI Act | Shaping Europe's Digital Future". Accessed 11 January 2025. https://digital-strategy.ec.europa.eu/en/policies/regulatory-framework-ai

Fricker, M. 2007. *Epistemic Injustice: Power and the Ethics of Knowing*. Oxford: Oxford University Press.

Frank, Arthur W. 2013. *The Wounded Storyteller: Body, Illness, and Ethics, Second Edition*. Chicago: University of Chicago Press. https://press.uchicago.edu/ucp/books/book/chicago/W/bo14674212.html

Future of Life Institute. 2024. "High-Level Summary of the AI Act | EU Artificial Intelligence Act". *EU Artificial Intelligence Act*, 27 February 2024. https://artificialintelligenceact.eu/high-level-summary/

HLEG. 2019. "Ethics Guidelines for Trustworthy AI". Brussels: European Commision.

Kara, Mahmut Alpertunga. 2023. "Is It Possible to Allocate Life? Triage, Ageism, and Narrative Identity". *The New Bioethics: A Multidisciplinary Journal of Biotechnology and the Body* 29 (4): 322–39. https://doi.org/10.1080/20502877.2023.2261735

Johnson, G. M. 2021. "Algorithmic bias: On the implicit biases of social technology". *Synthese* 198 (10): 9941–9961. https://doi.org/10.1007/s11229-020-02696-y

Jones, James H. 1993. *Bad Blood: The Tuskegee Syphilis Experiment. New and expanded edition*. Cambridge: Free Press.

Kittay, Eva Feder. 2019. *Learning from My Daughter: The Value and Care of Disabled Minds*. New York: Oxford University Press.

Lurie, P., and S. M. Wolfe. 1997. "Unethical Trials of Interventions to Reduce Perinatal Transmission of the Human Immunodeficiency Virus in Developing Countries". *The New England Journal of Medicine* 337 (12): 853–56. https://doi.org/10.1056/NEJM199709183371212

Obermeyer, Ziad, Brian Powers, Christine Vogeli, and Sendhil Mullainathan. 2019. "Dissecting racial bias in an algorithm used to manage the health of populations". *Science* 366 (6464): 447–453. https://doi.org/10.1126/science.aax2342

Regulation (EU) 2024/1689 of the European Parliament and of the Council of 13 June 2024 laying down harmonised rules on artificial intelligence and amending Regulations (EC) No 300/2008, (EU) No 167/2013, (EU) No 168/2013, (EU) 2018/858, (EU) 2018/1139 and (EU) 2019/2144 and Directives 2014/90/EU, (EU) 2016/797 and (EU) 2020/1828 (Artificial Intelligence Act).

Robertson, John A. 1983. "Procreative Liberty and the Control of Conception, Pregnancy, and Childbirth". *Virginia Law Review* 69 (3): 405–64. https://doi.org/10.2307/1072766

Ross, W. D. 2009. *The Right and the Good*. British Moral Philosophers. Oxford: Clarendon Press.

Rozée, Virginie, Anna De Bayas Sanchez, Michaela Fuller, María López-Toribio, Juan A. Ramón-Soria, Jose Miguel Carrasco, Kristien Hens, Joke Struyf, Francisco Guell, and Manon Vialle. 2024. "Reflecting sex, social class and race inequalities in reproduction? Study of the gender representations conveyed by 38 fertility centre websites in 8 European countries". *Reproductive Health* 21 (150). https://doi.org/10.1186/s12978-024-01890-2

Savulescu, J. 2001. "Procreative Beneficence: Why We Should Select the Best Children". *Bioethics* 15 (5–6): 413–26. https://doi.org/10.1111/1467-8519.00251

Scully, J. L. 2008. *Disability Bioethics: Moral Bodies, Moral Difference*. Lanham: Rowman & Littlefield.

———. 2022. "Being Disabled and Contemplating Disabled Children". In *The Disability Bioethics Reader*, edited by Joel Michael Reynolds and Christine Wieseler. London: Routledge.

———. 2023. "Feminist Bioethics". In *The Stanford Encyclopedia of Philosophy*, edited by Edward N. Zalta and Uri Nodelman. Stanford: Stanford University. https://plato.stanford.edu/archives/fall2023/entries/feminist-bioethics/

Silliman, Jael, Marlene Gerber Fried, Loretta Ross, and Elena Gutiérrez. 2004. *Undivided Rights: Women of Color Organize for Reproductive Justice*. Chicago: Haymarket Books.

Skloot, Rebecca. 2010. *The Immortal Life of Henrietta Lacks*. Crown Publishing Group.

Soled, Derek. 2021. "Public Health Nudges: Weighing Individual Liberty and Population Health Benefits". *Journal of Medical Ethics* 47 (11): 756–60. https://doi.org/10.1136/medethics-2020-106077

Tengland, Per Anders. 2016. "Behavior Change or Empowerment: On the Ethics of Health-Promotion Goals". *Health Care Analysis* 24 (1): 24–46. https://doi.org/10.1007/s10728-013-0265-0

Tolchin, Benjamin, Sarah C. Hull, and Katherine Kraschel. 2021. "Triage and Justice in an Unjust Pandemic: Ethical Allocation of Scarce Medical Resources in the Setting of Racial and Socioeconomic Disparities". *Journal of Medical Ethics* 47 (3): 200–202. https://doi.org/10.1136/medethics-2020-106457

trail. 2024. "EU AI Act: Risk-Classifications of the AI Regulation". Accessed 11 January 2025. https://www.trail-ml.com/blog/eu-ai-act-how-risk-is-classified

Van Assche, Kristof, Serge Gutwirth, and Sigrid Sterckx. 2013. "Protecting Dignitary Interests of Biobank Research Participants: Lessons from Havasupai Tribe v Arizona Board of Regents". *Law, Innovation and Technology* 5 (1): 54–84. https://doi.org/10.5235/17579961.5.1.54

Vertommen, Sigrid, Bronwyn Parry, and Michal Nahman. 2022. "Global Fertility Chains and the Colonial Present of Assisted Reproductive Technologies". *Catalyst: Feminism, Theory, Technoscience* 8 (1): 1–17. http://doi.org/10.28968/cftt.v8i1.37920

5. Animal Ethics and Animal Experimentation

Animal ethics

Discussion: Do you think it is permissible to eat other-than-human animals? What arguments for and against eating animals can you come up with?

Humans and other-than-humans

What are our duties and rights towards other-than-human animals? Should we eat them? Should we use them in animal experiments? Discussions around these topics often lead to heated debates. The way we view animals, whether it relates to their place in our lives or the cultural context we find ourselves in, influences our perspectives on practices such as consuming meat. We differentiate between animals that we classify as food and those we classify as companions. Why do we draw such distinctions? Is it even acceptable to draw moral distinctions between humans and other-than-human animals? Often, discussions surrounding animal experiments or meat eating resort to lifeboat scenarios. In such scenarios, people are asked to weigh the lives of a human vs an other-than-human animal. In the lifeboat case, it is then assumed that, of course, if there is only a place for three in a lifeboat, and there are three humans and a dog, the dog will have to go. However, such oversimplified scenarios fail to encompass the complexities of real ethical considerations. Indeed, because many intuit that the dog would have to go, it is automatically extrapolated that it is acceptable to eat meat or to engage in animal experimentation. However, we tend to forget that the lifeboat is an exceptional situation, one that is probably never going to happen. In fact, it has nothing to do with the complexities of the real-life ethics of animal experimentation or meat eating.

Before we delve into ethical questions regarding animals, more specifically animal experimentation, it is interesting to look at how views on human relations with other-than-human animals have evolved historically. What follows focuses on Western philosophy. However, we must not forget that much can be gained through engaging with other-than-Western approaches that have not taken the supremacy of humankind at face value.

In *On the Soul*, Aristotle (1984) describes the existence of different types of souls, each associated with specific properties and functions. According to Aristotle, plants possess the vegetative soul, which allows them to grow, reproduce, and nourish themselves. Animals, in addition to the vegetative soul, possess a sensitive soul, enabling them to perceive their surroundings, experience sensations, and engage in basic forms of cognition. Humans, according to Aristotle, possess vegetative and sensitive souls as well, but what sets them apart is the rational soul. This rational soul endows humans with the unique ability to engage in higher-order thinking and reason and possess intellectual capacities, distinguishing them as the pinnacle of the natural world. At the same time, there is a continuum from plant to human being. In Christian thinking in medieval times, this idea that humans are above plants and other-than-humans took over. At the same time, there was also the idea of God, who is above humans. Indeed, human beings are created in God's image, but at the same time, they share lower functions with other-than-human animals. What we should aspire to, however, is to be more like God and less like our animal brethren. Animals, moreover, are created to serve human beings. This is the idea of *separation*. We must strive to be more god-like, and too much engagement with animals is frowned upon or even morally suspect. This idea still survives in the modern day. Expending lavish amounts of money or love on one's companion animals is often seen as untoward, as if this is somehow misguided.

In early modern philosophy, René Descartes firmly separated animals and humans (1972). What we have in common with them is our body, which is machine-like, an automaton. Only human beings have souls and can feel and think. This is the origin of modern animal experimentation. As animals have the same 'machinery' as human beings, but no soul, they can be cut open and experimented on at will. At the same time, it is rumoured that Descartes had a dog, Monsieur Gnat, that he doted on. This is a fine example to demonstrate that even if we are rationally convinced that animals are mere automata, this conviction is overruled by the relation we have with them.

Charles Darwin put human beings right back in the continuum of the tree of life (1871). We are animals, we have animals as ancestors. This does not automatically mean that human beings fall from their pedestal. Some people would argue that human beings are at the top of the tree of life, that they are the acme of evolution. Human beings are the most evolved, the most superior. Others argue that there is no such hierarchy in evolution. We are animals amongst other animals.

Animal ethics

What rights and duties do we have towardsother-than-human animals? What rights do they have? To answer these questions, we can look back at the moral theories we saw in Chapter 2. Jeremy Bentham, one of the arch-fathers of utilitarianism, had a hedonistic view of what is good. For him, what is good is pleasure, and what is bad is suffering. This means that for him, all creatures that can suffer count, and should be included in moral reflection.

> The day may come, when the rest of the animal creation may acquire those rights which never could have been withholden from them but by the hand of tyranny. The French have already discovered that the blackness of skin is no reason why a human being should be abandoned without redress to the caprice of a tormentor. It may come one day to be recognized, that the number of legs, the villosity of the skin, or the termination of the os sacrum, are reasons equally insufficient for abandoning a sensitive being to the same fate. What else is it that should trace the insuperable line? Is it the faculty of reason, or perhaps, the faculty for discourse? [...] the question is not, Can they reason? nor, Can they talk? but, Can they suffer? Why should the law refuse its protection to any sensitive being? [...] The time will come when humanity will extend its mantle over everything which breathes. (Bentham, 1789, chapter 17)

Peter Singer, a contemporary ethicist, has taken up Bentham's idea in his seminal work *Animal Liberation (Singer, 2002)*, as we have seen in the chapter on environmental ethics. He is often associated with animal rights, but in fact, rights are deontological concepts. They are somewhat inalienable. In a utilitarian approach, interests can be weighed against one another. And here lies the difficulty. How do we weigh the suffering of a more-than-human animal against the joy or the suffering of a human being? Which animals can suffer? How can we know that for sure? This is called *'the problem of other minds'*. How do you know that others suffer or do not suffer as much as us? In animal experimentation, it is often assumed that we need more knowledge, hence more experiments, to find out whether certain animals can experience pain. But maybe it is better to assume that animals can suffer.

Different ethical approaches yield contrasting perspectives on the treatment of animals. When adopting a deontological approach instead of utilitarianism, discussions revolve around rights, duties, and personhood. Immanuel Kant, whom you might remember from our discussion on deontology, regarded rationality as the defining characteristic of being deserving of respect. Consequently, animals lacking rationality were considered mere means rather than beings to be given intrinsic value. However, proponents of the animal rights movement, such as Tom Regan (Regan, 2004), propose viewing animals as 'subjects of a life'. Although animals may lack rationality, they possess futures, life goals, and interests in survival—traits that warrant respect, duties, and rights. This perspective diverges from utilitarianism, as it emphasizes inherent entitlements rather than outcomes, regarding animals as inherently valuable.

There are individuals who challenge the assertion of animal rights and argue from a contractarian standpoint. Contractarians suggest that morality is founded on a social contract, wherein individuals agree to abide by certain rules. According to this perspective, animals do not belong to the moral community since they are not part of this contractual agreement. Only beings who willingly enter into the contract are considered moral beings and possess rights. Consequently, animals are excluded from these rights. However, contractarian approaches face criticism due to the 'argument from marginal cases'. This argument states that there is no morally relevant property to distinguish all humans from all other-than-human animals, so marginal cases (newborns, people in a persistent vegetative state, etc.) and animals should be treated alike. If you argue that it is acceptable to experiment on animals because they lack a certain characteristic, such as rationality, you run into a problem because many humans might also lack this characteristic. Philosopher R. G. Frey suggests that while there may be important differences between typical adults and animals, the ethical justification for conducting research on certain humans, such as those in a vegetative state, might be even stronger than for research on other-than-human animals, if we consider having higher cognitive capacities as morally relevant (1988). Hence, if research on such humans is deemed immoral, there is no morally justified basis for conducting research on sentient creatures who meet or exceed the conditions that protect the marginal cases. Although this argument might be considered ableist, it does show that it is nigh impossible to find arguments to include all human beings in the moral realm and exclude all non-humans.

Others have argued that there is a characteristic that binds all human beings and separates them from other-than-human beings, and that is the fact that they belong to the human species. Hence, there is a symbolic value assigned to human species membership that needs no further proof. For many ethicists, however, this is a form of speciesism. Speciesism is the (unwarranted) assignment of different values, rights, or special considerations to individuals based on their species membership, without further motivation. The term was first used by Richard Ryder of the Oxford Group in 1970, and popularized by Peter Singer (1975). Speciesism is analogous to sexism and racism. These authors all challenge the notion that arbitrary factors such as species, sex, or race should determine an individual's moral status or entitlement to equal consideration and respect.

Thinking about topics such as meat eating and animal experimentation is difficult. It is hard to defend our current practice of using animals as means to our ends based on their rationality or lack of self-consciousness. Every week, new findings regarding the intricate lives of animals are published. In the words of Mary Midgley, animals matter. In her book *Animals and Why They Matter* (2007), Midgley argues against lifeboat arguments that vet the lives of humans vs against those of other-than-humans. She says that there is no homogenous group of 'animals', but each species must be considered separately. Moreover, it is important to recognize that we have relationships with certain animals and not with others. Generalized moral theories and principles will only take us so far in thinking about what we owe them.

Animal experimentation

Introduction

Other-than-human animals are being used in scientific contexts for various purposes, including fundamental research, testing, education and training, the creation and maintenance of genetically altered animal models, and the use of organs or tissues. Under European law, an animal experiment or 'procedure' has been defined as "any use, invasive or non-invasive, of an animal for experimental or other scientific purposes, with known or unknown outcome, or educational purposes, which may cause the animal a level of pain, suffering, distress or lasting harm equivalent to, or higher than, that caused by the introduction of a needle in accordance with good veterinary practice" (Directive 2010/63/EU, Art. 3, L277/39). In 2021, out of all animals used in animal experiments in the EU and Norway, more than 96% were rodents, fish, birds, and rabbits. 40.9% of animals were used for basic research, 31.2% for translational and applied research, and 22.5% for regulatory use and routine production (ALURES database). Directive 2010/63/EU is only applicable to live animals; the killing of animals solely for the use of their organs or tissues is not considered and, thus, not recorded as an animal experiment.

> Inflammatory bowel disease (IBD) refers to conditions characterized by chronic inflammation of the gastrointestinal tract. In the long term, patients with IBD are at risk of developing colorectal cancer. A researcher is interested in the molecular mechanisms behind this development, from chronic inflammation to cancer, and suspects a specific protein X to be involved. If this is true, protein X could be a potential target for therapeutic interventions. To test this hypothesis of protein X's involvement, the researcher will use a genetically altered mouse model where the gene for protein X is deleted. For a period of twelve weeks, twenty genetically altered mice and a control group of twenty mice without genetic alterations will be subjected to a toxic substance that induces gut inflammation followed by colorectal tumour formation. The mice experience increasing discomfort and pain (diarrhea, bloody stool, etc.) over the course of the treatment. After the treatment, the mice will be euthanized and their colons will be dissected to evaluate tumour formations.
>
> - Do you think the researcher can use these mice to test their hypothesis? Why do you think this?
> - Would your opinion change if the researcher used worms, chickens, or pigs for the experiment?
> - Does the purpose of the research matter?
> - Would you set any conditions for the research to be allowed?

> - Would your initial opinion change if you knew that the experiment did not lead to any conclusive results in the end?

Two questions on animal experimentation

Animal experimentation has been the subject of many debates. Questions and criticisms can generally be divided into two topics: (1) the scientific utility and validity of animal use, and (2) the ethical permissibility of animal use. The first topic concerns the question of whether the scientific use of other-than-human animals leads to valid, useful, and relevant results. The second debated question is whether it is permissible for humans to subject other-than-human animals to pain, suffering, and death to achieve these results. Of course, the question of moral justification is also related to the question of scientific justification. If the use of animals in research does not result in any useful knowledge that could not be gained through other approaches, it will be harder to morally justify harm caused to them. However, even if animal experimentation can be justified scientifically, the question of ethical justification still remains.

The scientific utility and validity of animal experiments

The general scientific rationale for using other-than-human animals in research and testing is based on the need to advance scientific knowledge, from which humans and other animals can benefit. Animal models are considered valuable in research on disease mechanisms, the development of therapeutic interventions, and testing the safety and toxicity of various substances or interventions. The use of other-than-human animal models is generally based on their anatomical and physiological similarities to humans. Mammals, in particular, are seen as informative models for human anatomy and (patho)physiology because of their close evolutionary distance from humans. Human diseases—such as infectious diseases, cancer, and epilepsy—also affect other-than-human animals, so studying disease mechanisms in those animals might be informative for the medical knowledge of humans. The mouse models for studying rheumatoid arthritis (McNamee et al., 2015) and the rhesus monkey model for polio vaccine development (Curtis, 2004) are some examples of successful animal models for human disease. Although alternatives are being developed, animal models are generally still deemed necessary for the investigation and evaluation of system- and organism-level physiological functions and interactions in biomedical research.

However, some critics of animal experimentation question the scientific utility and validity of (some) animal disease models and, specifically, the transferability of results from the other-than-human animal model to humans. Indeed, despite biological similarities, interspecies differences may hinder the extrapolation of results from animal studies to the human context. Critics often refer to the fact that more than

90% of drug candidates identified in preclinical studies (including animal studies) have failed in clinical trials (Dowden and Munro, 2019). For example, HIV vaccines that seemed effective in chimpanzees have failed in subsequent human clinical trials, and no animal HIV model has been able to capture all features of human HIV-1 infection (Hatziioannou and Evans, 2012; Policicchio et al., 2016). In addition to the risk of incorrectly identifying drug candidates for human diseases, potential drugs or interventions for humans may be disregarded due to their failure in animal models.

Based on these criticisms of the scientific utility or validity of other-than-human animal models, some opponents of animal experimentation argue that all animal models are not sufficiently useful and should be replaced with other approaches, such as cell or tissue cultures, post-mortem research on humans, and computer simulations, which are deemed more reliable for research on human conditions. However, not all critics of the scientific validity of animal studies take such an *abolitionist* stance. Some do not deny that animal models may be useful in some cases, but assert that the predicative value of animal models is overstated and that other approaches may be more reliable for obtaining results relevant to humans. Others acknowledge that animal models may not always offer the most suitable approach, but argue that other modelling approaches—such as *in vitro* human cell or tissue models—also have limited transferability and predictive value. They argue that, although critical reflection is needed, animal research can still be scientifically valid or even necessary for research or testing.

In response to the debate, the Nuffield Council on Bioethics has issued the following conclusion on the scientific validity of animal and research testing:

> We concluded that continuities in the form of behavioural, anatomical, physiological, neurological, biochemical and pharmacological similarities provide sufficient grounds for the hypothesis that animals can be useful models to study specific aspects of biological processes in humans, and to examine the effects of therapeutic and other interventions. [...] 'the scientific validity of animal experiments is a condition capable of being fulfilled, but has to be judged case by case and subjected to detailed critical evaluation'. (Nuffield Council of Bioethics, 2005, p. 178)

In this conclusion, the Council affirms that animal experiments can be scientifically valid, while also acknowledging the limits of animal models of human disease and pointing to the need to carry out a critical evaluation of the study design and decide on the validity of an animal model on a case-by-case basis.

The ethical permissibility of animal experiments

The second big question in the debate on animal experimentation concerns the morality of humans using other-than-human animals for research, and, in particular, the permissibility of subjecting the animals to pain, suffering, and death for research. Various arguments for and against have been posed. Disagreements often come down to two aspects: (1) the moral status of humans and other-than-human animals, and (2) the reasoning and conclusion on the acceptability of animal research. However, there are also other points of debate, such as the extent of suffering other-than-human animals experience and whether or not we have a duty to alleviate and/or prevent it.

The moral status of humans and other-than-human animals

Assessments of the morality of humans using animals for research and testing often begin by considering the moral status of other-than-human animals: do other-than-human animals have moral status or moral importance, and how does it relate to the moral status of humans? Generally, there are three positions on this debate: (1) the clear-line view, (2) the moral equality view, and (3) the moral sliding scale view.

According to the clear-line view, there is a categorical moral division between humans and other-than-human animals. This view is based on the assumption that there is some morally relevant, specific property that is unique to or possessed by humans and that all other-than-human animals lack. This specific property is then considered vital for clearly assigning higher moral importance to humans.

The moral equality view posits no such categorical moral distinction between humans and all (or some) other-than-human animals. It claims that biological species classifications, as such, are insufficient for delineating humans as having higher moral importance than all other-than-human animals. As a result, some proponents of this view (e.g. Richard Ryder, Peter Singer) consider it 'speciesist' to assign higher moral importance to humans on the basis of species membership. Drawing analogies with sexism and racism, the concept of speciesism is characterized by an unjustified bias in favour of the interests of one's own species. According to this view of moral equality, humans and (some) other animals should be considered moral equals.

In between these two extremes, the moral sliding scale view argues against the clear dividing line for moral importance between humans and all other-than-human animals, opting for a scale ranking moral importance. According to this view, one or more specific features can be used to decide on a hierarchy of moral importance with, for example, humans at the top, followed by primates, rodents, zebrafish, fruit flies, and single-cell organisms. The morally relevant properties can be biological—e.g. the scale of neurological complexity—or not biological—e.g. 'capacity to flourish', which we have already discussed in the chapter on environmental justice. Other examples of morally relevant properties can also be found in the chapter on environmental ethics.

It should be noted that these views on the moral status of animals do not give straightforward answers to the question of whether it is ethically permissible to experiment on animals. For some, ascribing lower moral status to animals would justify all animal experimentation (an 'anything goes' view), while for others, the higher moral status of humans includes a moral duty of stewardship, care, or compassion to 'lesser' beings. This challenge of stewardship is also relevant for the moral sliding scale view. Additionally, the acceptability of animal research may not solely depend on the specific characteristics of the animals used, but also on the experiment's welfare implications for the animals. The moral equality view also does not necessarily defend or oppose animal experimentation.

In relation to the different views on the moral status of humans and animals, we can ask what features of humans and animals could be relevant to the assignment of moral status and subsequent constraints on how animals or humans can be treated or used. Various morally relevant features have been proposed, among which sentience, higher cognitive capacities, capacity to flourish, sociability, and possession of life have been the most popular. These features have been used in defence of different views on the moral status of humans and other-than-human animals, either as a single overriding criterion or as part of a combination of criteria for deciding position on a moral hierarchy. In subsequent discussions on the acceptability of animal research, different weights are assigned to these morally relevant features—some of them are regarded as absolutely sufficient to constrain the use of an animal in research, whilst some must be balanced with other factors.

There are multiple approaches to the ethical consideration of animal use for research, which partly depend on the moral theories followed. The approaches are generally based on different views on morally relevant features and their normative consequences. Ethical considerations on animal experimentation are usually similar to or an extension of approaches to animal ethics in general (see earlier). Here we briefly consider consequentialist, deontologist, and hybrid approaches.

Consequentialist approaches

Consequentialist approaches involve weighing up consequences to determine the acceptability of an action. A consequentialist evaluation of a specific animal experiment may require the consideration of three questions. First, there is the question of how the goals of the research are valued. This question often comes down to evaluating the benefits of the specific research goal. It is also important to consider for whom those benefits apply and how speculative the gains might be. One issue with this approach is the difficulty in predicting the value of a certain type of research beforehand, especially in the case of basic research. There is some disagreement on whether only immediate benefits should be considered or whether there is also intrinsic value in contributing to the overall sum of scientific knowledge. Second, there is the question about *the degree of harm experienced by the animals*. This depends on the number of animals used, and their capacity to experience pain, suffering, distress, or other harms from being subjected to the experiment. It might also be relevant to consider harm experienced during breeding, transport, housing, and handling. However, the degree of harm experienced by animals is also difficult to assess and doing so requires approximations and definitions of harm. For example, there is some debate on whether prematurely ending an animal's life, even if it does not suffer any pain or distress from the action, should be considered as harm to the animal or not. Finally, consequentialists may question whether a *better balance between the costs and benefits* is possible. Consequentialism requires an optimization of the overall consequences. Each research project should investigate whether other approaches would produce a better balance between overall benefits and costs. Are there, for example, non-animal alternatives available to achieve the valuable research goal, for which the degree of harm would be significantly reduced? Additionally, if no non-animal alternatives are available, a better balance could be achieved by reducing the number of animals used and/or refining the severity of procedures in order to reduce the harm experienced.

One of the main difficulties of consequentialist approaches is determining whether benefits outweigh costs. Even if the benefits and costs of a certain type of research can be reasonably identified, the direct comparison of benefits and costs often proves quite difficult. What degree of animal suffering would be too much in relation to the predicted benefits of the research? Is it more important to minimize suffering or maximize benefits? How do we integrate the degree of uncertainty for envisioned consequences in this calculation? Is there a difference in moral weight between causing suffering to animals and alleviating suffering for humans? This balance also depends on the assumed view on the moral status of humans and other-than-human animals.

In *Animal Liberation* (2002), Peter Singer follows a consequentialist approach to the ethical permissibility of animal experimentation. He starts from a view of moral equality view in the sense that he does not consider species as a relevant moral factor for distinguishing between the moral status of humans and other-than-human animals. Moreover, he regards all beings capable of suffering as worthy of equal consideration of their interests. Thus, in the context of animal experiments, the interests of the other-than-human animals should also be taken into consideration for the overall cost-benefit evaluation. His position does not *de facto* reject all animal research, but it requires scientists to clearly demonstrate the benefits of the research in comparison to the suffering inflicted on the animals, as well as evaluation by a board including scientists and members of the animal welfare community.

Deontology

As mentioned earlier, deontological approaches to animal ethics are generally based on rights, duties, and personhood.

Immanuel Kant (1963) found that, in contrast to humans, all other-than-human animals lack rationality, which he considers a vital property for the assignment of moral status (*clear line view*). In a Kantian view, rationality is required in order to have inherent value, and to be deserving of respect or rights. Following this deontologist view in the strictest sense, we would have no moral obligations towards animals, meaning they can be used as mere means to an end, although Immanuel Kant himself admitted that we should not be unnecessarily cruel to animals. This 'anything goes' position would allow all other-than-human animals to be used for research, regardless of the benefits of the research provided it does not violate the rights of any humans.

The animal rights movement finds no categorical distinction between humans and (some) species of other-than-human animals (*moral equality view*), and, in contrast to Singer, argue from that position in deontological terms. Tom Regan (2004) focuses on the inherent rights of animals instead of considering their interests and suffering. He suggests that, just like humans, animals have inherent value as 'subjects of a life'. Hence, his conclusion is also based on deontological reasoning, although he draws a different conclusion than Kant's because his view on animals is different. Animals have rights because of their inherent value as 'subjects of a life' and thus need to be treated with respect, not as mere means to an end. As a result, Regan holds an abolitionist view on animal use in science. Every form of animal research—regardless of its potential benefits for humans or other animals—would violate the animal's rights, so we have a duty not to use animals for science.

Hybrid approach

The most common approach to ethics in animal experimentation is a hybrid approach based on the sliding scale of moral status. An animal's position on the scale depends on a number of defined morally relevant features, which impose limitations on how the animal may be treated for research. The morally relevant features include sentience (the capacity to feel pain and pleasure), higher cognitive capacities (self-consciousness, rational will, communication, tool use, having moral systems, etc.), the capacity to flourish (have interests/needs met) in a specific environment, sociability (relations with humans and other animals), and possession of life. Based on the possession of some morally relevant features, some animals are ruled out completely from use in research. The use of chimpanzees, for example, is generally prohibited because they possess higher cognitive capacities. Within those very strict limits, the hybrid approach allows costs and benefits to be weighed up to evaluate the use of animals. For example, using mice to test the safety of an important, frequently used chemical may be permitted if the test inflicts minimal pain on the mice.

So, among proponents of the hybrid approach, the ethical debate on animal experimentation boils down to disagreements on two questions: (1) 'what are the absolute limits?', and (2) 'how should morally relevant factors be weighed within the permitted limits?"

When ethically evaluating the use of a particular animal for research, the following questions at least should be considered:

1. *What are the goals of the research*? Are these immediate goals (e.g. translational research) or long-term (basic research)? Are the goals of the research valuable to pursue? What are the potential benefits of the research? Who would benefit?
2. *What is the probability of successfully achieving the benefits?* What is the probability of achieving the goals of the research? How likely are the

predicted benefits? Predictions can be made based on earlier experiments, literature, or similar research.

3. *Which animals are to be used?* Which animals would be most suitable to answer the research questions? The choice of animal model can be based on genetic, physiological, or structural similarity to humans, specific characteristics that facilitate research, the researcher's expertise, etc. These considerations are important to ensure the scientific validity of the animal use. Is the most suitable animal 'prohibited' from use for (this type of) research? If so, would another animal model be appropriate for the research?

4. *What is the effect of the research on the animals?* Do the research procedures cause harm or distress to the animals, and what is the severity of the harm/distress? What is the effect of breeding, housing, transportation, and handling for the research? Is the harm or distress experienced throughout the entire experiment or only at a specific time point?

5. Are there any alternatives? Is it possible to replace the animal model with a cell or tissue model, computer model, etc.? Would a 'lower' animal also be appropriate for answering the research questions? Would another animal experience less harm from the research environment? Would another experimental approach require fewer animals, or reduce the harm or distress experienced by the animals?

Most European legislation follows the hybrid approach to the ethical permissibility of animal research. Directive 2010/63/EU sets an absolute limitation on scientific animal use by prohibiting the use of great apes in procedures (for the EU definition of 'procedure', see introduction). Additionally, some research goals are also restricted, with an EU-wide ban on testing finished cosmetic products on animals (Regulation (EC) N° 1223/2009), and a Belgian ban on animal research for the development of tobacco products (Royal Decree of 28 October 2008).

Outside of those limitations, EU (and Belgian) legislation focuses on balancing the costs and benefits of animal use by stipulating replacement, reduction, and refinement (the three Rs). The three Rs were first published in *The Principles of Humane Experimental Technique* by Russel and Birch (1959) as an approach to improve the treatment of animals in research, as well as the quality of animal studies. In current practice, the three Rs should be used to minimize animal use as well as the potential harm, pain, or distress experienced by animals in scientific research.

- *Replacement*: The substitution of live, conscious higher animals with insentient material (e.g. cell or tissue cultures, in-silico models).
- *Reduction*: The reduction in number of animals used to obtain information of a given quantity and precision. This often requires statistical power analyses to determine the minimum number of animals required to gain sufficiently significant results.

- *Refinement*: Any decrease in the incidence or severity of inhumane procedures. This also includes using enrichment approaches to decrease stress or harm experienced by the animals during experimental procedures, breeding, housing, transport, and handling. Examples of enrichment approaches include providing nesting materials in mice cages (Olsson and Dahlborn, 2002), or tickling lab rats at regular intervals (Cloutier et al., 2015).

To promote replacement, reduction, and refinement in scientific animal use, the European Union Reference Laboratory for Alternatives to Animal Testing (EURL ECVAM) has been set up. This laboratory is tasked with validating methods that reduce, refine, or replace the use of animals for safety testing and efficacy/potency testing of chemicals, biologicals, and vaccines.

> **Discussion:** Can you think of situations or types of experiments which would be difficult to evaluate based on the hybrid approach and 3Rs? Think of conflicts between interests, concepts, principles, etc.

Numerous criticisms and remarks have been made on the current hybrid approach to ethically evaluating scientific animal use. Some criticisms, such as those from the animal rights movement, stem from differing moral theories followed and differing views on the moral status of animals. Other criticisms do not necessarily disagree in those regards but do point out some weaknesses or conflicts within this current approach.

In some cases, there are conflicts between the reduction and refinement requirements of the three Rs. To reduce the required sample size for sufficient statistical power, variability in measurements has to be reduced. However, reducing response variability may require experimental methodologies which cause more pain and distress to the animals. Is it better to use fewer animals but cause more severe harm to those animals, or to use more animals but subject them to less severe harm? The hybrid approach does not provide straightforward answers to these conflicts.

Some critics point to the difficulty and arbitrariness of comparing and weighing morally relevant factors on which distinctions between different species are made in the hierarchy of moral importance. Why do the higher cognitive capabilities of great apes justify a ban on their use in research, but not the sociability of beagles? Why are the restrictions set out by Directive 2010/63/EU applicable to the use of vertebrate animals and cephalopods, but not other invertebrates (*C. elegans*, fruit flies, etc.)?

Lastly, the 'humane killing' of animals to use their tissues and organs in research does not fall under the definition of 'procedure' or 'animal experiment' in European law. As a result, prematurely ending the life of an animal is considered insufficient harm to require the same restrictions as other harms such as pain or stress.

Conclusion

Arguments on the ethical acceptability of humans eating, using, or experimenting on other-than-human animals take various forms. In this chapter, we introduce animal ethics and begin to show the range of approaches to animal experimentation ethics. We highlight two ways in which arguments on the ethical permissibility of animal experimentation can differ: namely, the moral status of other-than-human animals and the moral theories on which the argumentation is built. Along these dimensions, we provide students and researchers with a brief overview of some arguments for and against animal experimentation. Finally, we also lay out the hybrid approach to animal experimentation ethics, which underpins current evaluation policies and EU regulation on the ethical acceptability of animal studies.

Bibliography

"28 OKTOBER 2008. - Koninklijk besluit tot wijziging van het koninklijk besluit van 30 november 2001 houdende verbod op sommige dierproeven voor wat betreft de uitvoering van dierproeven voor de ontwikkeling van tabaksproducten". https://etaamb.openjustice.be/nl/koninklijk-besluit-van-28-oktober-2008_n2008024478.html

Aristotle. *De Anima* (On the Soul). Translated by J. A. Smith, in The Complete Works of Aristotle, edited by Jonathan Barnes. Vol. I. Princeton: Princeton University Press, 1984.

Beauchamp, Tom L., and R. G. Frey, eds. 2011. *The Oxford Handbook of Animal Ethics*. Oxford Handbooks. Oxford: Oxford University Press.

Bentham, J. 1789. *An Introduction to the Principles of Morals and Legislation*. London: T. Payne.

Cloutier, Sylvie, Chelsea Baker, Kim Wahl, Jaak Panksepp, and Ruth C. Newberry. 2013. "Playful Handling as Social Enrichment for Individually- and Group-Housed Laboratory Rats". *Applied Animal Behaviour Science, Special Issue: Laboratory Animal Behaviour and Welfare* 143 (2): 85–95. https://doi.org/10.1016/j.applanim.2012.10.006

Curtis, Tom. 2004. "Monkeys, Viruses, and Vaccines". *The Lancet* 364 (9432): 407–8. https://doi.org/10.1016/S0140-6736(04)16746-9

Darwin, Charles. 1871. *The Descent of Man, and Selection in Relation to Sex*. 1st edition. London: John Murray.

Descartes, René. 1972. *Treatise on Man*. Translated by Thomas Steele Hall. Cambridge: Harvard University Press.

Dowden, Helen, and Jamie Munro. 2019. "Trends in Clinical Success Rates and Therapeutic Focus". *Nature Reviews Drug Discovery* 18 (7): 495–96. https://doi.org/10.1038/d41573-019-00074-z

European Commission. n.d. "ALURES - ANIMAL USE REPORTING - EU SYSTEM". Accessed 11 July 2024. https://webgate.ec.europa.eu/envdataportal/content/alures/section1_number-of-animals.html

——— 2009. "Regulation (EC) No 1223/2009 of the European Parliament and of the Council of 30 November 2009 on Cosmetic Products". November. https://health.ec.europa.eu/system/files/2016-11/cosmetic_1223_2009_regulation_en_0.pdf.

―――― 2010. "Directive 2010/63/EU of the European Parliament and of the council of 22 September 2010 and of the council of 22 September 2010 on the protection of animals used for scientific purposes". http://eur-lex.europa.eu/LexUriServ/LexUriServ.do?uri=OJ:L:2010:276:0033:0079:eN:PDF.

Frey, R. G. 1988. "Moral Standing, The Value of Lives, and Speciesism". *Between the Species* 4 (3): 191–201. https://doi.org/10.15368/bts.1988v4n3.8

Hatziioannou, Theodora, and David T. Evans. 2012. "Animal Models for HIV/AIDS Research". *Nature Reviews Microbiology* 10 (12): 852–67. https://doi.org/10.1038/nrmicro2911

Kant, Immanuel. 1963. *Lectures on Ethics*. Translated by Louis Infield. New York: Harper & Row.

McNamee, Kay, Richard Williams, and Michael Seed. 2015. "Animal Models of Rheumatoid Arthritis: How Informative Are They?". *European Journal of Pharmacology* 759: 278–86. https://doi.org/10.1016/j.ejphar.2015.03.047

Midgley, Mary. 2007. *Animals and Why They Matter*. Athens: University of Georgia Press. https://ugapress.org/book/9780820320410/animals-and-why-they-matter

Olsson, I. Anna S., and Kristina Dahlborn. 2002. "Improving Housing Conditions for Laboratory Mice: A Review of 'Environmental Enrichment'". *Laboratory Animals* 36 (3): 243–70. https://doi.org/10.1258/0023677702320162379

Policicchio, Benjamin B., Ivona Pandrea, and Cristian Apetrei. 2016. "Animal Models for HIV Cure Research". *Frontiers in Immunology* 7: 12. https://doi.org/10.3389/fimmu.2016.00012

Regan, Tom. 2004. *The Case for Animal Rights*. 2nd edition. Berkely: University of California Press.

Reza Khorramizadeh, M., and Farshid Saadat. 2020. "Animal Models for Human Disease". *Animal Biotechnology*, 153–71. https://doi.org/10.1016/B978-0-12-811710-1.00008-2

Russell, William Moy Stratton, Rex Leonard Burch, and Charles Westley Hume. 1959. *The Principles of Humane Experimental Technique*. Vol. CCXXXVIII. London: Methuen.

Singer, Peter. 2002. *Animal Liberation*. 2nd edition. New York: Ecco.

"The Ethics of Research Involving Animals". 2005. London: Nuffield Council on Bioethics. https://www.nuffieldbioethics.org/publications/animal-research

"The Internet Classics Archive | On the Soul by Aristotle". n.d. Accessed 11 July 2024. http://classics.mit.edu/Aristotle/soul.html

Van Norman, Gail A. 2019. "Limitations of Animal Studies for Predicting Toxicity in Clinical Trials". *JACC: Basic to Translational Science* 4 (7): 845–54. https://doi.org/10.1016/j.jacbts.2019.10.008

6. Epigenetics

Introductory remarks

Why would we include an introduction to the ethics of epigenetics in an introductory bioethics textbook? Is epigenetics not a highly technical specialization of biology, hardly accessible to undergraduate students coming to bioethics from other disciplines? We hope to explain the basics of epigenetics in a simplified, understandable fashion in this chapter, because an ethical discussion of epigenetics may be illuminating for bioethics students in at least two ways:

- Epigenetics is an interesting case study to discuss the social and ethical implications of scientific progress in a domain that shows an intricate connection between our body and our environment, or our biology and our biography.
- It allows us to demonstrate that scientific research projects are never value-neutral: in the case of epigenetics, findings can be employed to bolster a wide variety of claims with regards to individual or societal responsibilities, and the priorities of the research also reveal what societies value and want to prevent.

Introduction to epigenetics

The modern term *epigenetics* has multiple related meanings. Firstly, it denotes heritable—via mitosis and/or meiosis—changes in gene function without changes in DNA sequence. Secondly, epigenetics refers to the study of those processes and mechanisms and their implications for biological functioning. To avoid misunderstandings, this introduction sometimes uses terms such as 'epigenetic mechanisms' when referring to the first sense of the word and phrases such as 'epigenetic research' and 'epigenetic knowledge' in the context of the second sense. This introduction will provide some scientific background on aspects of epigenetics that are relevant to ethical discussions.

Epigeneticists do not study changes in DNA itself but rather mechanisms that influence how and when genes—which are stretches of DNA bases—are expressed in an organism. Epigenetic mechanisms can affect the transcription and translation of genes in various ways. Two important processes are histone modification and DNA methylation.

- *Histone modification*: The histone is a kind of spool made of proteins around which the genomic DNA is wrapped to save space. The complex of the DNA and the histone proteins is called chromatin. How tightly the DNA is wrapped around the histone influences how easily the DNA can be accessed and thus copied. The more tightly packed it is, the less gene expression is possible. Tightly-packed and thus less accessible parts of the chromatin are called heterochromatin. The more readable parts are called the euchromatin—genes can only be expressed when they are located here.

- *DNA methylation*: This epigenetic mechanism involves the addition of a methyl group to a DNA molecule. This does not change the DNA itself, but it does influence whether certain parts of it can be read and transcribed. We can think of DNA methylation as a process to 'silence' genes by making them inaccessible.

By regulating gene expression, epigenetic processes influence cell types and tissues phenotypes (observable characteristics), function, and developmental state. Firstly, epigenetic programming is responsible for the differentiation of stem cells into specialized cells, providing them with the 'memory' of their differentiated identity. This explains how all cells in an organism contain the same DNA while still performing a wide variety of functions. Each of the ~400 tissues of the human body, for example, has a different epigenome (i.e. a different set of epigenetic modifications), whereas all the cells share a single genome, usually.

In addition to its function in cell differentiation, epigenetics also has other functions throughout the lifetime of an organism. One way in which our epigenome changes is the 'epigenetic drift' associated with aging. In general, more epigenetic changes indicates older age, meaning that epigenetic marks can be seen as biomarkers of aging. However, our epigenome changes mostly in response to environmental stimuli, which are most relevant for ethical perspectives on epigenetics. Mechanisms such as DNA methylation can be triggered by environmental factors, both stemming from within the body and from the outside environment. Crudely put, this means that the material and psychosocial circumstances of our body—our diets, the quality of the air we breathe, and the stress we experience—can impact epigenetic mechanisms. This is why epigenetic mechanisms are often treated as missing links between our lifestyle/environment and our physical/mental health.

Metaphor: Playing the piano

Perhaps a metaphor integrating some of the process outlined above will be helpful at this point. Epigenetics can be understood by thinking of a musician such as a piano player (Raz, Pontarotti, and Weitzman 2019). The piano player interprets or decodes the musical score when they want to play a composition. The score is analogous to the encoded message of the DNA: multiple musicians might follow the same score, just like multiple nuclei contain the same DNA. How the piece is performed, however, depends on the interpretation of the piano player—and how the DNA is expressed depends on the epigenetic mechanisms at work. Even if they follow the same musical score, two pianists may perform the piece in completely different ways. They may choose to add notations to the sheet music indicating the speed and dynamics they want to use in specific sections (as violinists might add 'bow notations'), or the emphasis they want to put on some notes. Such annotations are usually made with a pencil so that the pianist can still erase or re-write them. Epigenetic methylation patterns on the DNA are also dynamic to a certain extent, which means that they can change over time. The interpretation of each musician, in turn, depends on environmental factors and is thus subject to change. A pianist may alter their playing style of the same piece depending on whether they play it for their family at home or in a big concert hall. Similarly, epigenetic signals can be triggered by environmental factors.

Epigenetic information can be regarded as another layer beyond genomic information, enriching but also challenging more traditional understandings of genetics. For example, it challenges the 'central dogma' of molecular biology which assumes that genetic information flows only in one direction, when DNA is transcribed into RNA which is in turn translated into proteins that determine the phenotype. Epigenetics shows that the interface between genes and their environment is much more complex.

Epigenetic inheritance

Can the epigenetic marks that someone accumulates due to environmental exposure and lifestyle be transmitted to subsequent generations? This question has been intensely discussed and has led to much speculation and ethical theorizing in the past two decades. Most epigenetic programming is rewritten or reset between generations, but there is increasing evidence that this is not always the case. When considering the transmission of epigenetic marks between generations, we need to distinguish between *transgenerational* and *intergenerational* effects.

Intergenerational epigenetic inheritance refers to epigenetic marks in offspring that are the result of *direct* exposure of their parental germline (sex cells) to environmental stressors. This means that intergenerational inheritance is limited to the first generation of male offspring (i.e. children) and the first and second generations of female offspring. Transmission to the first generation of offspring means that epigenetic marks are passed on from one generation to the next (i.e. from parent to child). The second generation of female offspring is also included under intergenerational inheritance because oocytes (egg cells) are already present in a female foetus in the womb. This means that environmental triggers during pregnancy can directly affect not only a first, but also a second generation of offspring.

A famous example of intergenerational epigenetic inheritance occurred during the Dutch Hunger Winter famine of 1944–1945 (Heijmans et al., 2008). The children of mothers who experienced this famine during their pregnancy were found six decades later to have reduced DNA methylation of the imprinted IFG2 gene, which is associated with the risk of metabolic diseases. These and other findings lend empirical support to the hypothesis that early-life environmental conditions can cause epigenetic changes in humans that persist throughout their lives. Public discourse and research often focuses on maternal factors. However, epigenetics shows that paternal factors and postnatal exposures in later life can play a role in offspring health, in addition to influences *in utero*. We will come back to this in the final section of this chapter.

Transgenerational epigenetic inheritance is more contested. It denotes the *indirect* transmission of epigenetic information that is passed on to gametes without alteration of the DNA sequence. As was explained earlier, direct epigenetic inheritance (i.e. intergenerational inheritance) pertains to the passing on of epigenetic information to the first generation of male and female offspring and the second generation of female offspring (since her sex cells were already exposed to external influences in the womb of her grandmother). This means that we can only speak of transgenerational inheritance if the epigenetic effects of the first generation's environmental exposures are still present in the second generation of male offspring or the third generation (i.e. great-grandchildren) of female offspring. So far, most transgenerational epigenetic effects have been discovered in plants and other-than-human animals such as rats and mice. For example, researchers working with mice have found third-generation epigenetic effects of maternal diet as well as social stress levels, although others argue that multigenerational inheritance of methylation patterns in mice is an exception rather than the rule (Dunn and Bale, 2011; Kazachenka et al., 2018). A study of *C. elegans* worms by Adam Klosin and colleagues also had impressive results (Klosin et al., 2017). They genetically modified these worms to glow when exposed to a warm environment. The worms not only started to glow more when the temperature was raised, but they also retained their intense glow when researchers lowered the temperature again. Moreover, even seven generations further down the line, glowing offspring were born. If five generations of *C. elegans* worms were kept warm, this characteristic was passed on to fourteen generations.

Unfortunately, in research on human inheritance, it is virtually impossible to exclude potential confounding elements such as changes in utero and postnatal effects. It is hard to distinguish 'real' epigenetic inheritance from cultural inheritance or reconstruction of the environmental context resulting in the same experiences or health problems in offspring. Still, some studies indicate that transgenerational epigenetic inheritance is possible, albeit limited, in humans. Studying historical data of cohorts in Överkalix, researchers found correlations between grandpaternal food supply and the mortality rate of the following two generations, their children and grandchildren (Kaati et al., 2017). Because no molecular data were available, no epigenetic links could be proven. Pembrey and colleagues build on these findings to evidence sex-specific male transgenerational inheritance in humans (Pembrey et al., 2006). In a longitudinal study of men in an area around Bristol, they found transgenerational effects of smoking before puberty on the growth of future male offspring. Specifically, early paternal smoking (before puberty) was associated with a greater body mass index (BMI) in their sons. The researchers posit DNA methylation as a potential mechanism behind those links between the acquired epigenetic traits of a generation and the epigenetic marks present in the next generations.

Diseases, conditions, and cures

The following list is a selection of epigenetic research on various human diseases and conditions. It is not exhaustive, but it is intended to give you an idea of the broad scope of epigenetics research.

- Exposure to *stress* in the womb or during early childhood has been associated with epigenetically mediated adverse health effects. For example, childhood maltreatment might trigger long-lasting epigenetic marks, contributing to PTSD in adult life. Researchers have found that children of survivors of the 9/11 attack in the USA who were pregnant at the time seem more vulnerable to PTSD and behavioural issues (e.g. Jablonka, 2016). Others argue that epigenetic processes might link the antenatal mood of the mother (e.g. maternal depression) to how infants will respond to new situations (Oberlander et al., 2008).

- As is well known, *air pollution* has numerous harmful effects on health. Emerging data indicate that exposure to air pollution modulates epigenetic marks (Rider and Carlsten 2019). These changes might in turn influence inflammation risk and exacerbate the risk of developing lung diseases.

- It is well known that *lead* is a common neurotoxic pollutant that disproportionally affects the health of children. Evidence of the epigenetic basis of the effects of lead is increasing (Wang et al., 2020; Senut et al. 2012)

- The epigenetic mechanisms behind the development *of metabolic conditions* such as type 2 diabetes, diabetic kidney disease (DKD), and obesity are increasingly well-documented. Like stress, obesity has been posited not merely as a health outcome but also as a causal factor in epigenetics. Paternal prepubescent obesity has been associated with diminished lung function and asthma in adult offspring (Lønnebotn et al., 2022).

- *Neuroepigeneticists* investigate how epigenetic regulation plays a crucial role in the development and functioning of our brain. Conditions for which epigenetic regulatory mechanisms have been suggested include Parkinson's, Huntington's, schizophrenia, epilepsy, Rett syndrome, and depression. Much research seems to be particularly geared towards a better etiological understanding of neurodevelopmental conditions such as Tourette's syndrome, ADHD, and autism. However, there is still much uncertainty about the concrete causative evidence that might be implicated in the development of such conditions.

Epigenetic changes seem to be more readily reversible than genetic ones. This reversibility holds promising potential for epigenetic therapies for diseases, since epigenetic marks such as methylation patterns can be seen as targets for medical interventions and treatments.

Many clinical research efforts in this domain are directed toward the treatment of *cancers*. Cancer cells are often characterized by epigenetic drifts, and many tumours are associated with epigenetic reprogramming. While some studies investigate the possibility of epigenetic interventions in general, others focus on specific types of cancer such as breast cancer and prostate cancer. There are many 'epidrugs' for cancers in clinical trials, but research on epidrugs for other conditions is also very prolific. Recent projects have aimed to target conditions such as Covid-19, hypercholesterolemia, neurodegenerative diseases, autoimmune diseases such as chronic kidney disease, and depression.

Ethics of epigenetics

Now that we have a basic understanding of epigenetics, we can start thinking about ethical issues concerning the research field and its findings. There are several aspects to epigenetic findings that we can take into account when thinking about the ethics of epigenetics:

1. *Influence of environment:* As we saw earlier, you can think of epigenetic mechanisms as a kind of missing link between a person's lifestyle and environmental influences on the one hand and that person's physical and mental health on the other. In other words, epigenetics makes us think about the link between our biology and our biography.

2. *Heritability:* Some epigenetic markers or changes that occur under the influence of environmental factors over a lifetime seem to be passed on to subsequent generations (e.g. the earlier example of the Dutch Hunger Winter). Another important insight is that epigenetic changes in prospective parents can affect offspring even if they occur before conception. This means that the behaviour and lifestyle of people who may not even be thinking

about having children at all yet can have an impact on the health of those future children.

3. *Reversibility:* Epigenetic changes are relatively dynamic. As we saw, this ensures that a lot of promising progress is being made in the field of epigenetic treatments for all kinds of diseases and disorders. But does this also have ethical implications? For example, if you can reverse the development of some conditions, do you have the responsibility to do so? Is it less bad to contribute to someone's ill health (e.g. by polluting a neighbourhood) if the effects may be reversible to some extent? And what about the distribution of cures or medicine—how can epigenetic treatments be distributed fairly?

4. *Uncertainty:* Finally, the actions of epigenetic mechanisms are quite unpredictable. We know that certain factors can affect those mechanisms, but we are far from knowing everything about all of them, and the question is whether that will ever be possible. After all, all kinds of environmental factors also interact. At best, it seems possible to make predictions that are accurate to some extent and to talk about 'increased odds' or a certain predisposition. Does this uncertainty make a difference in our ethical reflection, when compared to the judgements we would make if we were certain of all the mechanisms and effects involved? Can we hold people responsible for actions that may or may not have a certain effect?

Ethical issues in the literature

What, then, are some normative issues that require a closer look in light of epigenetic findings? In their literature review, Dupras, Saulnier, and Joly (2019) identify nine areas of discussion at the crossroads of epigenetics, law, and society: traditional nature-nurture dichotomy; embodiment or 'biologization' of the social; public health and other preventive strategies; reproduction, parenting and the family; political theory; legal proceedings; the risk of stigmatization, discrimination or eugenics; privacy protection; and knowledge translation. Other widely discussed issues include environmental justice, the need for bioethical approaches that integrate concern for both the environment and medicine, and ethical, legal and social issues of epigenetics research in the context of personalized medicine.

Privacy

Ensuring that the privacy of patients and research participants is respected should always be a priority for researchers. But insights from epigenetics make the safe collection and storage of health data even more urgent. For example, by gathering information about epigenetic markers (such as the DNA methylation we encountered before) it becomes very easy to identify people based on anonymous donor material.

Moreover, perhaps some agents might be interested in using these data for non-medical purposes. Information about lifestyle behaviours, whether you smoke or drink for example, might be interesting to insurance companies or employers. This is why researchers have already been calling for strict regulations and laws that forbid genetic or epigenetic discrimination. These concerns are becoming increasingly relevant with the growing availability of commercial direct-to-consumer epigenetic tests.

Exposing inequality

Epigenetics seems to provide new possibilities to map existing injustices as well as the distribution of social and environmental factors that may have a long-term, detrimental effect on people's health. It is well known that not everyone is exposed equally to external harm. Just look at recent examples in Belgium such as the 3M PFAS scandal—where local residents were found to have elevated concentrations of the toxic chemical PFAS in their bodies—and the Umicore factory in Hoboken—where lead pollution has a big impact on small children (see Case 1 below). Many people also believe that such unequal exposure is, at least in some cases, problematic or unjust. Epigenetic markers could potentially indicate who exactly is hit the most by existing inequalities. And in cases for which we already have an indication—we do not need epigenetics to tell us that people living close to highways suffer more from air pollution—epigenetics might provide new information about the impact on individuals and future generations.

Responsibility

Most normative discussions on the ethics of epigenetics are conducted in terms of distributing responsibility. For example, we can ask ourselves 'who is morally responsible for the health of current and future generations in the context of epigenetics'?

When we ask *who* can or should be responsible, we can distinguish between *individual, shared,* and *collective responsibility*. The idea of collective responsibility is that a group has certain characteristics that make it a moral agent that can be held responsible. This idea is not uncontested. However, most authors believe that some organizations with a clear structure—such as corporations, governmental organizations, medical organizations, or NGOs—can be said to have some kind of responsibility in the context of epigenetics.

What dimensions of responsibility do we have in mind? We often think about moral responsibility in retrospective or *backward-looking* terms. We blame people for bad outcomes, or shower them with compliments because they did something with a good outcome. But it is also interesting to think about future-oriented or *forward-looking* responsibility: how we can attribute responsibility and take responsibility with an eye on what we want to achieve or avoid in the future? When we think about our responsibility for the health of future generations, responsibility can function as a way to distribute moral labour. If we know what state of affairs we want to achieve, we can start thinking about a desirable allocation of the tasks that should be performed in order to get there.

Gunnar Björnsson and Bengt Brülde (2017) created a list of factors which might help to identify moral agents and inform responsibility distribution. This could be useful for distributing responsibility for the epigenetic health of future generations. This is their list:

1. *Capacity and cost*: The individual or group that has most capacity to produce a good outcome is responsible for doing so.
2. *Retrospective and causal responsibility*: There is still some connection between what an agent did in the past and who needs to take up responsibility now, even though this is perhaps not as important as other considerations.
3. *Benefiting*: Benefiting from another's help or benefiting from harm, injustice, or danger to others can make an agent more responsible, to reciprocate for example.
4. *Promises, contracts, and agreements*: If an agent promises to deal with a problem, the agent takes on responsibility by doing so.
5. *Laws and norms*: Laws can prescribe our behaviour and can give direction to and limit our ascribed responsibility.
6. *Roles and special relationships*: We can have more responsibility towards people we have a close connection to.

Three cases

Below are three cases related to epigenetics. They all have the following features in common: (1) they link environmental influences to health outcomes, and (2) they invite us to ask ethical questions. You may use these cases to practice your ethical reflections, applying concepts, theories, and ideas.

Case 1: Hoboken

Hoboken—a district of the Belgian city of Antwerp—is home to a factory site of Umicore, one of the world's largest refiners of precious metal. The factory plant is surrounded by a residential area that was constructed over the twentieth century. Emissions of lead, cadmium, and arsenic by the factory have been contributing to widespread health problems in children from the surrounding area for decades (Pano, 2021). Despite efforts that have greatly reduced both the emissions and their impact, lead levels in the blood of children living in the neighbourhood continue to exceed the standards set by public health agencies. Epigenetic mechanisms may contribute to lead-induced health effects in children, such as behavioural issues and problems with developing gross motor skills (Senut et al., 2012; Wang et al., 2020).

Parents worry about the health of their children and often feel guilty about living in the vicinity of the polluting factory. In a documentary, Esther—a mother of two children with very high lead values in their blood—expresses her worries as follows: "I want my child to be able to be himself. If he is good at something, it should be possible to stay that way. And if he is not as good at something, that should not become worse. I do not want external factors I have no control over, such as the factory, to interfere [...] Stay away from my child, is what I think" (Pano, 2021, my translation). As the Flemish report agency Pano succinctly puts it, parents seem to be given a choice between "kuisen of verhuizen"—cleaning or moving.

Case 2: Mexico City

Since 1993, the ELEMENT (Early Life Exposure in Mexico to Environmental Toxicants) project has investigated the impact of environmental factors such as toxins and sugars on mother-child pairs in various neighbourhoods of Mexico City. Anthropologist Elizabeth Roberts collaborated with this project. First, her fellow researchers found that eating from traditional lead-glazed plates—which are said to make the food taste sweeter—was the surest predictor of high lead levels in mothers and children. The exposure to lead is both gendered—because it is women who prepare the food and inherit the plates from their (grand) mothers—and cultural, because the plates connect their users to a rural past.

Additionally, the high consumption of sweets and sugary soda is said to be an important factor in the high rates of obesity and diabetes in poorer neighbourhoods of Mexico City. Soda is almost as cheap as bottled water and is more reliably available than tap water. Inhabitants know that soda and sweets can lead to ill health, but "in Moctezuma sharing soda, liquid-food, filled with sugar, is love" (Roberts, 2015, p. 248). Because it performs important social roles, campaigns exhorting individuals (primarily mothers) to stop providing soda to their children have little effect.

Finally, there is a penetrating smell caused by "a narrow stream of dam runoff, filled with *aguas negras* (untreated sewage) and garbage" (Roberts, 2015, p. 592) in these neighbourhoods. In rainy seasons the dam often overflows, leaving the walls of the cement houses with salmonella, *E. coli*, and faecal enterococcus.

The effects of these various exposures on inhabitants of this neighbourhood are not only direct, but can also be inherited through epigenetic mechanisms.

> ### Case 3: Farah and Alex
>
> Farah is a postdoc researcher at a prestigious university. She loves her job and considers being an academic an important part of her identity. At the same time, various elements of her job are causing her quite some stress. When Farah gets pregnant, she makes a conscious decision to continue working her stressful job, even though she is aware of the potential influence of the accumulated stress on her offspring. Ten years later, her child Alex receives a diagnosis of ADHD after experiencing some difficulties in home and school settings. Although he sometimes continues to struggle with aspects of his ADHD, throughout his teenage years Alex starts to consider ADHD an integral part of his identity that he would not want to change.
>
> Suppose Alex learns about studies that imply a link between stress during pregnancy and ADHD in offspring through epigenetic mechanisms (e.g. Bock et al., 2017). Maybe when Alex is in college, he talks with his mother to learn more about the decisions she made before and during her pregnancy. He may want to learn more about the decisions she made, and the circumstances that perhaps constrained them. What might his reaction be when he finds out more about her reasons for continuing her stressful work? Should he blame her, or could understanding her situation instead help to strengthen their bond?

In all three cases, epigenetics might be a part of the puzzle of biologically explaining the link between environment and health (although epigenetic links are not the *only* causal connection at play in any of these cases). A second feature these cases have in common is that they may invite us to ask ethical questions. What, if anything, should be done about the situations in Hoboken and Mexico City? Who, if anyone, is to blame for negative health outcomes in inhabitants of those places? What duties do (future) parents such as Esther or Farah have towards their offspring concerning their health? What does it mean, or what should it mean, to say that we want children to be healthy? How does that relate to wanting to protect them from the harmful effects of pollution? What is the role of scientists, policymakers, and public health institutions? Do social injustices exacerbate health disparities, and if so, what does that imply for our moral evaluation? All of these questions are, in one way or another, related to an overarching question of responsibility: who is responsible for what with regard to whom?

One finding, many ethical and political claims

For each of the ethical aspects of epigenetic research mentioned above, it is possible to formulate many different viewpoints and moral claims.

Take 'reproduction, parenting, and the family' as an example. We know that epigenetics deepens and extends our knowledge about the impact of parental behaviour as well as environmental factors on the development of foetuses and young children alike. However, there are a myriad of ways in which this knowledge might be translated into moral or political claims. Especially in popular science communication, perinatal influences are often cited to inflate the individual responsibility of parents (and mothers in particular). However, as philosopher Daniela Cutas points out, epigenetic knowledge might also help us to see that the category of biological parenthood may need to be broadened. It seems safe to assume that everyone who is closely involved in raising a child influences their environment and experiences and also modulates their molecular biology in doing so. She suggests that those people may not have parental or procreative responsibilities, but based on their epigenetic contributions they might have a biological 'responsibility for shaping'. Seen from that perspective, epigenetic knowledge production "brings closer together or altogether blurs the margins between parental, non-parental, primary, secondary, individual and collective responsibilities for children" (Cutas, 2024, p. 107). The final section of this chapter will delve deeper into this theme.

Consider responsibility for environmental pollution as another example. As was explained in the introduction to epigenetics, epigeneticists can point out with increasing accuracy which environmental stimuli trigger epigenetic mechanisms that contribute to adverse health outcomes. But how can we translate this into moral claims or political action? This is a contentious matter. Who bears responsibility for certain kinds of pollution, or for the impact on public health? How do we distribute responsibility among many actors in a complex network of interactions? Big factories may be clear culprits in some cases, but governmental organizations allowing them to operate in harmful ways may also have responsibilities here. What about the health of animals and whole ecosystems? And how do we qualify 'adverse health effects' or 'epigenetic harm'? In the case of Farah and Alex, for example, it becomes clear that Alex does not think of his ADHD as something harmful, but rather as a valuable part of his identity. Although neurodiversity theory is currently spreading awareness about the problems with a deficit approach to conditions such as ADHD and autism, much epigenetic research unfortunately still defines those conditions only in terms of deficits.

We may want to ascribe responsibilities to involved individuals and collectives on a variety of grounds and moral principles. As we already saw more generally in earlier chapters, no moral theory can provide a clear-cut answer that points in just one direction here. In short, epigenetic research never holds straightforward implications for healthcare and society. One reason for this is that epigenetic mechanisms in and of themselves are often not sufficient for a disadvantageous outcome. Instead, they always interact with largely pre-existing social, economic, and environmental factors and (dis) advantages. Epigenetic findings alone cannot tell us when a situation is unjust, nor do they provide specific ways to combat situations we do characterize as unjust.

In addition, we need to keep in mind that knowledge production in science is never a morally or politically neutral process either, and epigenetics is no exception. Scientists, research institutions, and funding agencies all have their own moral, political, and scientific values that impact their research, and choices made in the process of knowledge creation are always context-dependent. This is not unique to epigenetics; but as a science that has quickly gained a lot of broader public attention, it does make a very good example. This is also not a problem that needs to be avoided or mitigated. Rather, it is important that everyone involved is explicit about their values and considers how those influence their work.

Epigenetic knowledge itself thus cannot simply be regarded as either a burden or a blessing, but at best as a "double-edged sword" (Meloni, 2016). Epigenetic knowledge can be applied to morality and politics in a variety of ways depending on the values, commitments, priorities, and biases of the person applying it. Thus, it is important that researchers explicitly state which values underlie their research claims wherever possible. One way to start thinking about this may be to distinguish between (1) the normative lens, (2) moral theories, and (3) considerations of various practical and normative aspects that might play a role. A *normative lens* is an overarching perspective or paradigm that characterizes the commitments of a researcher, such as a commitment to egalitarianism, or an intersectional feminist approach. *Moral theories*, such as those discussed in earlier chapters, can also guide one's judgement in certain directions, although they usually leave it open to various conclusions. Which *practical and normative aspects* of a situation you find relevant for the ethical judgement of a particular case depends on your normative lens(es) and the ethical theory you are working with. Your moral and political views may also influence the relative weight or importance you attach to each aspect.

Parental responsibility in epigenetics

The final section of this chapter zooms in on a specific ethical debate in the context of epigenetics: the responsibility of (prospective) parents for the health of their offspring. As we saw earlier, findings in intergenerational epigenetics give rise to a 'temporal expansion' of normative discussions about parental responsibility. This section is intended to show how ethicists working on epigenetics can take very different approaches based on the same research findings and societal context of their research. Broadly, we can distinguish between *cautionary* and *emancipatory* approaches.

Cautionary approaches

The lifestyles, behaviours, circumstances, and exposures of people who are planning to have a child, or are already pregnant, are indeed subject to intense normative scrutiny in both scientific and popular discourse, but is this fair or just?

Most of the literature on the potential ethical and social implications of epigenetic discoveries for procreation and parenthood takes a cautionary approach (Dupras, Saulnier, and Joly, 2019). Articles can generally be categorized as criticizing one or both of two tendencies they observe in the scientific and popular discourse on epigenetics:

(1) increased responsibilization of individual (prospective) parents, which does not take the wider social context into account; and (2) disproportionate focus on maternal (as opposed to paternal) behaviour before and during pregnancy.

Individual and collective responsibilities

Most scholars and philosophers working on the ethics of epigenetics have seen it as their task to criticize, either explicitly or implicitly, a scientific and societal discourse that excessively attributes responsibility to individual people—particularly mothers—for epigenetic alterations in offspring. They point out that expecting individual (prospective) parents to prevent disease or suboptimal epigenetic transmission in their offspring by minimizing every possible risk factor seems to ignore the extent to which exposures, diets, and stressors are shaped by social, economic, and political forces. In this context, scholars often emphasize the importance of collective responsibility. As we saw before, Daniela Cutas argues that the environmental aspect of epigenetic mechanisms implies that all agents who causally contribute to a child's environment being a certain way might together bear some collective responsibility for the child's wellbeing.

Another concern is the personal and private nature of individual and familial decisions about childbearing and raising children. Although dealing with the inequities that shape the lives of individuals and their children requires societal change, it is far from clear to what extent the state should be allowed to interfere.

Finally, even if we attribute collective responsibility for the epigenetics of future generations rather than holding individual parents solely responsible, it is unclear what should be done about this. Excessive blaming of individuals is a potential downside of accounts which heighten individual responsibility. On the other hand, increased social pressure and state interference might be a downside of an approach that puts too much emphasis on collective responsibility. For example, the line between preventing harm and optimizing or enhancing an outcome is not at all easy to draw, especially in the context of parental responsibility. Epigenetic findings might be employed to intensify societal pressure on individual parents (and especially women) to have healthy children, thereby "maximizing human capital and productivity" (Wastell and White, 2017, p. 178). Some commentators even worry about the risk of 'epi-eugenics' through "increased social pressure on prospective parents to undergo preconception and prenatal testing for epigenetic alterations" (Juengst et al., 2014, p. 428).

Maternal and paternal influences and responsibilities

Pregnancies, and thus women's bodies and behaviours, seem to have become the main target of intervention suggested in epigenetic literature. While epigenetics expands the temporal window of potential influence, this overemphasis on maternal influence itself on the health of a foetus, baby, or child is nothing new.

There seems to be a growing consensus among commentators that the overemphasis on maternal influences in epigenetic risk messaging is unfair because it risks ascribing excessive blame to women. In an influential paper, Richardson and colleagues compellingly demonstrate that narratives about epigenetic findings risk perpetuating "a long history of society blaming mothers for the ill health of their children" (Richardson et al., 2014, p. 131). They give examples from the media such as panic around Fetal Alcohol Spectrum Disorder (FASD) and 'crack babies' in the USA, and the popularity of theories about 'refrigerator mothers' whose 'cold' parenting style supposedly caused autism in their children. They warn us that although the scientific findings underpinning these societal blaming practices are often rather moot or proven to be plain wrong, women still experience the blame to this day.

Thus, researchers should be aware of existing biases and moralizing tendencies in society when they share their findings, in order to minimize the risk that their findings inspire unfair blaming practices. However, the problem seems to be situated on a more fundamental level than the biased language in science communication. The disproportionate attention and resources science has been directing toward maternal influences, specifically in the perinatal period, should also be critically questioned. As we saw earlier in this chapter, ethical considerations also play a role in the construction of epigenetic knowledge itself.

The focus on maternal influences seems to be a continuation of the centuries-old "bewitching idea that the environment in which you are gestated leaves a permanent imprint on you and your future descendants" (Richardson, 2021, p. 1). However, epigenetics offers an opportunity to strike a new balance in parental responsibility between contributors, because it shows that other influences besides *in utero* influences also play a role in offspring health.

In recent years, epigenetics researchers seem to have heeded calls to research paternal influences as well—e.g. by creating a POHaD paradigm (paternal originals of health and disease) that researches the impact of paternal lifestyle and exposure and their impact on, for example, sperm quality (Mayes et al., 2022). However, venturing into this area requires some caution. These new findings could reduce the burden of responsibility currently placed on mothers, but researchers should also be careful not to reconstruct the stigmatizing and blaming tendencies of discourse about maternal influences in the discourse about paternal influences.

Emancipatory approaches

We can conclude that most existing work on the ethics of epigenetics points out the dangers of employing this knowledge in such a way that overburdens (prospective) parents or blames them unfairly or disproportionally. Although such warnings are important and necessary, there might also be more positive or emancipatory ways of thinking about new developments in epigenetics.

We will now look at the examples of empowerment, procreative autonomy, and benefits to the parent-child relationship.

Empowerment and procreative autonomy

A first possible way to think positively about epigenetic knowledge in an unequal society is to see it as a tool in striving toward the *empowerment* of individual citizens and communities. There is no consensus about the definition of empowerment. It is usually understood as the process of enhancing people's capacities to control the determinants of their own quality of life ('empowering people'), and/or as the state that results from this process ('empowered people').

The use of 'empowerment' in health care is a good example of how a concept can be used to serve various ethical or political goals. For example, Luca Chiapperino and Giuseppe Testa are critical of how the concept of empowerment is often used (Chiapperino and Testa, 2016). They observe that epigenetic knowledge is frequently employed as the basis for a neoliberal project of individualizing responsibility for health. Language of empowerment can be used to serve this project that seeks to transfer responsibility for health from the state to individual citizens and expects this move to make the healthcare system more economically sustainable or even profitable.

However, Chiapperino and Testa do not rule out the potential use of empowerment in emancipatory discourse. They refer to a more radical history of the concept, for example in the tradition of liberatory pedagogy. They argue that epigenetic knowledge can be empowering in that it shows people how social factors and environmental exposures can affect their health and that of their offspring. What an empowerment discourse should be mindful of, then, is that people also need to be sufficiently free from financial, social, and material constraints to act on this knowledge.

Such an emancipatory project could be served by a sufficiently refined concept of *procreative autonomy*. That is the right of people to decide whether, when, and under which circumstances to procreate. A related concept is parental autonomy, which involves the rights of people to parent their children as they see fit. How autonomous an agent is depends not only on their capacity for self-governance but also on the extent to which they are socially and politically free to make decisions that impact their own life. Epigenetics is relevant to both parental and procreative autonomy. Perhaps epigenetic knowledge can empower vulnerable (potential) parents-to-be to make informed decisions.

Note that although procreative autonomy can benefit a (future) child, this is not necessarily the case. When we emphasize the right of future parents to make choices about procreation and pregnancy, we need to acknowledge that those choices need not always be good for the foetus' or future child's health. With limited state intervention or nudging, the actions of parents may well lead to worse health outcomes for future children or go against their interests more broadly. However, some room for making bad choices may need to be allowed in order to protect people from far-reaching public involvement in the private sphere, which could be even more harmful.

Parent-child relationship

Take another look at the case of Farah and Alex earlier in this chapter. Might there be any value for the child or the parent-child relationship in having shared knowledge of the lives of parents before they conceived? Or before they were even thinking of conceiving, when their experiences may have impacted the child's biological make-up nonetheless? If we answer 'yes' to this question, we can consider whether epigenetic knowledge has a role to play here. Epigenetic knowledge can perhaps play a contextualizing role, affirming that aspects of the child's health or personality are not isolated from the actions, behaviours, or exposures of their parents in the past. Consider an example related to environmental pollution:

> *Jenn and her parents*: Two people who intend to have children together have both grown up in a poor neighbourhood close to a polluting factory. They are aware of this pollution and its potential health effects on themselves and their future offspring. Although this is far from easy, they manage to move to another part of their city with relatively clean air. There, they conceive, and their daughter Jenn is born. However, their epigenetic marks from having lived in the polluted neighbourhood may have been inherited by Jenn to some extent.

Would it be valuable for Jenn to know this? And if so, how might she react?

Jenn might be thankful that her parents decided to move away from a place that they were very attached to for her sake. She might gain a better sense of appreciation for their considerations (although it is not unthinkable that she might also feel guilty for being the reason they made such a drastic and costly change). Moreover, knowledge about epigenetic mechanisms might help Jenn understand why she is more prone than others to certain conditions such as asthma. Conversations about the ways in which social determinants of health—conditions in the social and physical environments of people that influence health outcomes throughout their life course—have affected both Jenn and her parents may lead them to a sense of mutual understanding.

In short, the power of epigenetic knowledge might lie in helping people to integrate their biography and their biology. Another way to put this is that parent-child conversations on such topics help the child to create their own narrative identity: it can help parents, children, and families to tell stories about why they are who they are.

Moreover, epigenetic research can help to value the contributions different individuals make towards the upbringing of children. Lesbian mothers who gestate a child conceived with their partner's oocyte and donor sperm (Bower-Brown et al., 2024) and surrogates (Pande, 2009) often use epigenetic research as biological proof of their meaningful contribution to a child's being. In queer families, epigenetic effects are used to argue that kinship can be based on other biological connections than mere genetics.

Kinship is then seen as connections between people who care for each other every day, whatever the genetic bond between them. In the same way, epigenetic findings allow for a broader understanding of 'mothering' as a practice by multiple people: parents and other caretakers, for example day care staff or grandparents. Thus, epigenetic research not only enables multiple people (or the community at large) to be perceived as responsible for a child's well-being, but also enlarges the web of connections between them. By involving groups of people in kinship in a meaningful way, epigenetic knowledge has the emancipatory potential to queer existing kinship schemes.

Conclusion

In this chapter, we showed various ways in which researchers and students can engage in ethical discussions of developments in epigenetics. After a brief introduction to the scientific background of epigenetics, we formulated several ethically relevant aspects to epigenetic findings that we can take into account when we are considering the moral impact of such findings: the influence of the environment, heritability, unpredictability, and reversibility. We outlined ethical issues which are being discussed recurrently in bioethical literature on epigenetics, and presented readers with a few cases that invited them to ask ethical questions and practice moral reflections, applying concepts and theories. Finally, we discussed two particular issues in more detail: (1) how the case study of epigenetics demonstrates that scientific research projects are never value-neutral, and (2) how research findings can be employed in multiple ethical discourses in the specific debate on the responsibility of (prospective) parents for the health of their offspring.

Bibliography

Björnsson, Gunnar, and Bengt Brülde. 2017. "Normative Responsibilities: Structure and Sources". In *Parental Responsibility in the Context of Neuroscience and Genetics*, edited by Kristien Hens, Daniela Cutas, and Dorothee Horstkötter, 69, pp. 13–33. Cham: Springer International Publishing. https://doi.org/10.1007/978-3-319-42834-5_2

Bock, Joerg, S. Breuer, Gerd Poeggel, and Katharina Braun. 2017. "Early Life Stress Induces Attention-Deficit Hyperactivity Disorder (ADHD)-like Behavioral and Brain Metabolic Dysfunctions: Functional Imaging of Methylphenidate Treatment in a Novel Rodent Model". *Brain Structure and Function* 222 (2): 765–80. https://doi.org/10.1007/s00429-016-1244-7

Bower-Brown, Susie, Kate Shaw, Anja McConnachie, Vasanti Jadva, Kamal Ahuja, and Susan Golombok. 2024. "Biogenetic kinship in families formed via reciprocal IVF: 'It was [my partner]'s egg... but my blood flowed through her'". *Sociology* 58 (3): 735–752. https://doi.org/10.1177/00380385231212398

Chiapperino, Luca, and Giuseppe Testa. 2016. "The Epigenomic Self in Personalized Medicine: Between Responsibility and Empowerment". *The Sociological Review* 64 (1): 203–20. https://doi.org/10.1111/2059-7932.12021

Cutas, Daniela. 2024. "Epigenetics, Parenthood and Responsibility for Children". In *Epigenetics and Responsibility*, edited by Emma Moormann, Anna Smajdor, and Daniela Cutas, pp. 98–109. Bristol: Bristol University Press. https://doi.org/10.56687/9781529225440-007

Dunn, Gregory A., and Tracy L. Bale. 2011. "Maternal High-Fat Diet Effects on Third-Generation Female Body Size via the Paternal Lineage". *Endocrinology* 152 (6): 2228–36. https://doi.org/10.1210/en.2010-1461

Dupras, Charles, Katie Michelle Saulnier, and Yann Joly. 2019. "Epigenetics, Ethics, Law and Society: A Multidisciplinary Review of Descriptive, Instrumental, Dialectical and Reflexive Analyses". *Social Studies of Science* 49 (5): 785–810. https://doi.org/10.1177/0306312719866007

Heijmans, Bastiaan T., Elmar W. Tobi, Aryeh D. Stein, Hein Putter, Gerard J. Blauw, Ezra S. Susser, P. Eline Slagboom, and L. H. Lumey. 2008. "Persistent Epigenetic Differences Associated with Prenatal Exposure to Famine in Humans". *Proceedings of the National Academy of Sciences of the United States of America* 105 (44): 17046–49. https://doi.org/10.1073/pnas.0806560105

Hens, Kristien, Christina Stadlbauer, and Bart H. M. Vandeput. 2022. *Chance Encounters: A Bioethics for a Damaged Planet*. Cambridge: Open Book Publishers. https://doi.org/10.11647/obp.0320

Jablonka, Eva. 2016. "Cultural Epigenetics". *The Sociological Review* 64 (1): 42–60. https://doi.org/10.1111/2059-7932.12012

Juengst, Eric T., Jennifer R. Fishman, Michelle L. McGowan, and Richard A. Settersten. 2014. "Serving Epigenetics before Its Time". *Trends in Genetics* 30 (10): 427–29. https://doi.org/10.1016/j.tig.2014.08.001

Kaati, G, Lo Bygren, and S Edvinsson. 2002. "Cardiovascular and Diabetes Mortality Determined by Nutrition during Parents' and Grandparents' Slow Growth Period". *European Journal of Human Genetics* 10 (11): 682–88. https://doi.org/10.1038/sj.ejhg.5200859

Kazachenka, Anastasiya, Tessa M Bertozzi, Marcela K Sjoberg-Herrera, Nic Walker, Joseph Gardner, Richard Gunning, Elena Pahita, Sarah Adams, David Adams, and Anne C Ferguson-Smith. 2018. "Identification, Characterization, and Heritability of Murine Metastable Epialleles: Implications for Non-Genetic Inheritance". *Cell* 175 (5): 1259–71. https://doi.org/10.1016/j.cell.2018.11.017

Klosin, Adam, Eduard Casas, Cristina Hidalgo-Carcedo, Tanya Vavouri, and Ben Lehner. 2017. "Transgenerational Transmission of Environmental Information in C. Elegans". *Science* 356 (6335): 320–23. https://doi.org/10.1126/science.aah6412

Lønnebotn, Marianne, Lucia Calciano, Ane Johannessen, Deborah L. Jarvis, Michael J. Abramson, Bryndís Benediktsdóttir, Lennart Bråbäck, Karl A. Franklin, Raúl Godoy, Mathias Holm, Christer Janson, Nils O. Jõgi, Jorunn Kirkeleit, Andrei Malinovschi, Antonio Pereira-Vega, Vivi Schlünssen, Shyamali C. Dharmage, Simone Accordini, Francisco Gómez Real, and Cecilie Svanes. 2022. "Parental Prepuberty Overweight and Offspring Lung Function". *Nutrients* 14 (7): 1506. https://doi.org/10.3390/nu14071506

Mayes, Christopher, Elsher Lawson-Boyd, and Maurizio Meloni. 2022. "Situating the Father: Strengthening Interdisciplinary Collaborations between Sociology, History and the Emerging POHaD Paradigm". *Nutrients* 14 (19): 3884. https://doi.org/10.3390/nu14193884

Meloni, Maurizio. 2016. *Political Biology*. London: Palgrave Macmillan. https://doi.org/10.1057/9781137377722

Oberlander, Tim F., Joanne Weinberg, Michael Papsdorf, Ruth Grunau, Shaila Misri, and Angela M. Devlin. 2008. "Prenatal Exposure to Maternal Depression, Neonatal Methylation of Human Glucocorticoid Receptor Gene (NR3C1) and Infant Cortisol Stress Responses". *Epigenetics* 3 (2): 97–106. https://doi.org/10.4161/epi.3.2.6034.

Pande, A. 2009. "'It may be her eggs but it's my blood': Surrogates and everyday forms of kinship in India". *Qualitative Sociology* 32: 379–297. https://doi.org/10.1007/s11133-009-9138-0

Pembrey, Marcus E., Lars Olov Bygren, Gunnar Kaati, Sören Edvinsson, Kate Northstone, Michael Sjöström, Jean Golding, and ALSPAC Study Team. 2006. "Sex-Specific, Male-Line Transgenerational Responses in Humans". *European Journal of Human Genetics* 14 (2): 159–66. https://doi.org/10.1038/sj.ejhg.5201538.

Raz, Aviad, Gaëlle Pontarotti, and Jonathan B. Weitzman. 2019. "Epigenetic Metaphors: An Interdisciplinary Translation of Encoding and Decoding". *New Genetics and Society* 38 (3): 264–88. https://doi.org/10.1080/14636778.2019.1601009.

Richardson, Sarah S. 2021. *The Maternal Imprint: The Contested Science of Maternal-Fetal Effects*. Chicago: University of Chicago Press. https://doi.org/10.7208/chicago/9780226807072.

Richardson, Sarah S., Cynthia R. Daniels, Matthew W. Gillman, Janet Golden, Rebecca Kukla, Christopher Kuzawa, and Janet Rich-Edwards. 2014. "Society: Don't Blame the Mothers". *Nature* 512 (7513): 131–32. https://doi.org/10.1038/512131a

Rider, Christopher F, and Chris Carlsten. 2019. "Air Pollution and DNA Methylation: Effects of Exposure in Humans". *Clinical Epigenetics* 11 (1): 1–15. https://doi.org/10.1186/s13148-019-0713-2

Roberts, Elizabeth F. S. 2015. "Food Is Love: And so, What Then?" *BioSocieties* 10 (2): 247–52. https://doi.org/10.1057/biosoc.2015.18

Senut, Marie-Claude, Pablo Cingolani, Arko Sen, Adele Kruger, Asra Shaik, Helmut Hirsch, Steven T Suhr, and Douglas Ruden. 2012. "Epigenetics of Early-Life Lead Exposure and Effects on Brain Development". *Epigenomics* 4 (6): 665–74. https://doi.org/10.2217/epi.12.58

Wang, Tian, Jie Zhang, and Yi Xu. 2020. "Epigenetic Basis of Lead-Induced Neurological Disorders". *International Journal of Environmental Research and Public Health* 17 (13): 4878. https://doi.org/10.3390/ijerph17134878

Wastell, David, and Susan White. 2017. *Blinded by Science: The Social Implications of Epigenetics and Neuroscience*. Bristol/Chicago: Policy Press. https://doi.org/10.56687/9781447322368

7. Synthetic Biology

What is synthetic biology?

Synthetic Biology (SynBio) is a scientific field that has gained much prominence in popular media and scientific literature. In 1980, Barbara Hoom coined the term 'synthetic biology' to describe a specific class of genetically engineered bacteria using recombinant DNA technology. In 2000, Eric Kool and other speakers at the American Chemical Society annual meeting in San Francisco reintroduced the term to describe the synthesis of unnatural organic molecules that function in living systems.

SynBio refers to the building, modelling, designing, and fabricating of novel biological systems using customized gene components that result in artificially created genetic circuitry. This umbrella term covers a variety of research areas that can mainly be classified into two broad subfields. One *uses* unnatural molecules to produce a desired product from natural biology. The other seeks interchangeable parts from natural biology to assemble into systems that act unnaturally. The common goal for both subfields is to use interchangeable parts that can function independently to develop new systems that meet specific desired requirements. Identifying and creating such interchangeable parts or toolkits in the molecular world is the aim of SynBio.

SynBio has many applications. Drug discovery, reducing or improving our carbon footprint, and improving agriculture are three of its central goals. For example, in biomedicine, Synbio applications can accelerate molecular production, facilitate diagnosis through different health-monitoring systems using biochips or sensors to detect physiological changes, and revolutionize treatment procedures using new advances such as therapeutic nucleic acids, gene editing, and cell therapy, thereby enabling more accurate, targeted therapies. SynBio has also been suggested to have transformative potential for the agricultural and food industry. By programming plant activities and production, SynBio can help to improve the agricultural environment and enhance yield. Some examples of SynBio applications in agriculture are improving nitrogen fixation, reducing the use of synthetic fertilizers, improving the nutritional value of plants, aiding in soil remediation, and changing the production mode of chemical pesticides to biopesticides. As we will see later, however, while these developments can offer helpful solutions for longstanding agricultural problems, we should also recognize the associated risks of replacing traditional knowledge systems with scientific applications. Finding this balance is critical to ensure SynBio works as a boon in the agricultural domain, and for society at large.

SynBio research is also being used in biofuel production. SynBio uses different technologies to reprogram or create new microbes to aid in efficient biofuel production. SynBio not only could help improve the quality and efficiency of biofuel produced from traditional sources such as plants, but it could also enable the production of biofuels from non-traditional sources such as waste materials and novel microorganisms—by creating 'cell factories' capable of generating energy from the traditional and nontraditional feedstock. Creating new strains of novel microbes for biofuel production using natural or waste feedstocks, can enable the production of renewable and less toxic novel biofuels, thereby reducing the carbon footprint.

Conceptual issues in SynBio

When thinking about the ethics of SynBio, it is important that we first delve a bit deeper into some conceptual issues. First, the term SynBio covers much ground, which makes it difficult to create an 'ethics of Synbio'. For example, SynBio can refer to minimal genomes. The minimal genome is a concept that can be defined as the minimum set of genes sufficient for life to exist and propagate under nutrient-rich and stress-free conditions. It can also be defined as the gene set supporting life on a single cell culture in nutrient-rich media. It is thought that what makes up the minimal genome will depend on the environmental conditions that the organism inhabits. This minimal genome concept assumes that genomes can be reduced to a bare minimum, given that they contain many non-essential genes of limited or situational importance to the organism. Therefore, if a collection of all the essential genes were put together, a minimum genome could be created artificially in a stable environment. By adding more genes, the creation of an organism with desired properties is possible. The concept of a minimal genome arose from the observation that many genes are unnecessary for survival. To create a new organism, a scientist must determine the minimal genes required for metabolism and replication. This can be achieved by experimental and computational analysis of the biochemical pathways needed to carry out primary metabolic and reproductive functions. Some uses of the minimal genome are identifying genes essential for survival, thereby reducing the genetic complexity of synthetic strains to engineer—e.g. microbes designed to produce a desired product, or plants that survive in harsher conditions. These are just some examples of the many possibilities.

There is also the creation of *orthogonal biosystems*. The genetic information that all living systems require to function is stored, in coded form, in the sequence of the four types of sub-units that make up the long chains of DNA molecules. Researchers have been experimenting with various ways of modifying the system to carry the instructions for making types of protein unknown in nature. Even more radical is the synthesis and use of alternatives to DNA to create a new genetic material. Any alternative molecule would need properties comparable to DNA's—information storage, the ability to self-replicate, etc.—and should be able to function similarly. Living systems relying on an alternative of this kind might be unable to interact with conventional (DNA-based) life forms due to fundamental biochemical incompatibilities. Since the genetic circuits are designed using a distinct set of DNA bases and an alternative coding scheme, they can only be interpreted by organisms equipped with the corresponding molecular machinery As a result, these synthetic organisms would be unable to exchange genetic information with natural life forms. This process can potentially constitute a form of biological containment by preventing a created organism from surviving or interacting outside of its intended niche, which could have potential safety benefits (EASAC, 2011).

SynBio is also used to refer to *metabolic engineering*. This is the creation of new biosynthetic pathways to produce valuable materials that living organisms do not naturally create. It means engineering microbial or cell factories to produce the precursor to an end product or produce the product itself. Examples include the production of the anti-cancer drug taxol in the yeast *Saccharomyces cerevisiae* (Zhou, 2023), the creation of a precursor of spider silk using the bacterium *Salmonella typhimurium* (Widmaier, 2009), the manufacturing of second-generation biofuels in yeast (Basso, 2011), and the synthesis of hydrocortisone from glucose, again in yeast (Szczebara, 2003).

Regulatory circuits are another example of what is considered a SynBio application. The natural activity of cells is controlled by circuits of genes analogous to electronic circuits. So, new cell functions can be introduced by creating novel internal circuitry to alter their pattern of activity. Using well-understood genetic components that act as molecular switches, it should be possible to devise artificial gene networks. Linked together and implanted into natural systems, such networks could aid in control of what those systems do, when, and how frequently. Integrated into suitable cells, an artificial network might be used to sense and correct metabolic disturbances found in diabetes.

Science has, for a long time, drawn upon a variety of metaphors, including several from engineering. In *Metaphors We Live By* (1980), Lakoff and Johnson explain that language and metaphors shape how we understand the world. Scientific knowledge is structured at the primary level by certain concepts that shape our understanding of science. Metaphors are fundamental tools used to represent these concepts that structure scientific knowledge. Sometimes, those metaphors are not even evident because we use them unconsciously. For example, we say that the cell wall acts as a barrier. In this case, 'barrier' is a metaphor indicating that the cell wall is a separation or a protective layer. In other cases, the usage of metaphors is quite evident, e.g. we say DNA is the 'software of life', or we call genes 'codes' and bacteria 'chassis'. This also indicates the influence of computational or machine metaphors in biology. As the view of DNA as the 'software of life' became popular, scientists were driven by the idea that they might be able to direct cells like people program computers, but were confronted with the uncertainties and constraints of engineering in the cellular context.

All these different definitions and applications require different ethical considerations. Moreover, besides the question of what we are referring to when we talk about SynBio, the metaphors we use in science also influence our ethical reflection.

While viewing biology through the eyes of engineering, scientists are essentially trying to isolate each component of the living organism, understand its function, and rearrange them according to the desired final product. This is what engineers have been doing all along, but trying to apply the same principles in biology could create a sort of 'ethical puzzlement' for some. This is because fixing different blocks or units is how *machines* are created and understood, not life. Life has always been viewed as something created by nature, not by engineers. Thus, some have argued that this blurs the line between living organisms and machines. Organisms have a purpose: to self-generate, self-maintain, and perform their function. These are seen as intrinsic purposes. However, machines do not possess this; they possess extrinsic purposes determined by external agents. So, would it be right to view synthetic entities that can self-replicate, self-maintain, and evolve further as machines? The metaphor of 'living machines' in technologies like SynBio can create confusion among people who might have trouble with the idea of the sanctity of life now under human control. This could give rise to a slippery slope of problems, such as unattainable expectations of a utopian society, fear of playing God or overestimating human power, fear of unleashing a fierce creature, fear of eugenics, etc.

Metaphors are vital and inextricable in shaping our understanding; hence, they must be used responsibly. In the context of SynBio, metaphors play an even more critical role in shaping the emerging meaning of life and responsibility. They must be used responsibly because they are fundamental tools for thinking about and acting on the world. Metaphors matter, and they have direct and indirect ethical, legal, and social consequences, as well as political and economic ones. Metaphors can significantly impact science, policy, and public response in the context of synthetic biology.

Ethics of SynBio

Ethics in technology, simply put, refers to moral principles that govern how technologies should be utilized. SynBio is a powerful technology that allows us to design and create organisms/products to help us solve many current global problems, such as environmental damage or the lack of medicines. However, the ethical dilemma here is that we do not have complete control over our creation, and the outcome of our creation is highly unpredictable. In SynBio, our ability to create 'synthetic organisms with great power' and our inability to 'ultimately control' the actors involved raises ethical and social concerns. The actors here are not only the synthetic organisms we create but all the stakeholders involved in their creation—from the scientists in the lab to the funding agency and the governmental regulatory authorities.

SynBio is an example of a technology that could lead to *dual-use dilemmas*. Such dilemmas arise when scientific knowledge could be used in good and harmful ways, and the risk of harmful use is sufficiently high that it is no longer clear whether that knowledge should be pursued or disseminated. Besides the regular biosafety and biosecurity issues seen with genetic engineering, SynBio raises new concerns over the unpredictability and uncontrollability of creating new and novel entities. These 'human/lab-made' entities also raise various philosophical concerns challenging the current views and perceptions of life. Advocates of the technology state that it has great potential because its applications are so diverse—for example, it can be used to produce various bio drugs (or biologicals) and create tailor-made metabolic pathways for them, potentially transforming human life. At the same time, SynBio comes with its own set of ethical, legal, and social issues. SynBio has attracted the attention of philosophers, ethicists, anthropologists, and religious scholars, who warn about challenges surrounding the creation of de novo parts of biological processes and the potential unpredictability and uncontrollability of these components.

Ethicists have wondered about *the ethics of creating life* in synthetic biology. SynBio's ability to help create new entities from scratch in vivo has garnered attention and raised the eyebrows of many ethicists and philosophers. While the creation of life has always been seen as a power of nature or the divine, scientists can do the same in vivo, creating a slippery slope of concerns. The first is: are we humans taking up the role of the divine or are humans 'playing God?'. This is followed by fears of losing respect for the value of life. If life is eventually seen as something that can be manufactured in labs, would it lead to a loss of respect and humility toward the value of life? Synthetic biology could reduce life to just another product of industry, akin to other products. Are we, as humans, overstepping our boundaries in trying to protect nature?

When the creation of life shifts from 'nature' to 'labs', would life be seen as a technological production process? Once scientists create a new form of life or entity within these labs, what kind of 'moral status and moral values' can be attributed to it? Should they be considered alive because they fulfil the basic requirements such as metabolism, reproduction, etc., or should they be considered machines because they have been engineered in the lab? Do they possess intrinsic purposiveness (self-organizing, self-maintaining, and self-regenerating), or do they only possess extrinsic purposiveness (organized, assembled, and maintained by external agents), making them akin to machines? These questions arise because if we attribute 'moral status and moral values' to these entities, then—per some deontology theories—it would be wrong to use them for human benefit.

Besides uncertainties regarding the moral status of creating life and of the resulting life created, there are also social concerns when it comes to SynBio. These concerns can be broadly classified into three categories: knowledge-related, method-related, and application/distribution-related. Knowledge-related concerns are those related to knowledge creation and dissemination. The fear of misusing knowledge is a major concern in SynBio. The creation and dissemination of knowledge used to create new synthetic entities could also be misused. This concern raises questions about open sources and the sharing of knowledge. How do we draw boundaries around which forms of knowledge can be shared, to what extent, and with whom? There is the conflict of beneficence vs non-maleficence; while open science sources are essential to ensure the benefits of science reach all, if it ends up in the hands of biohackers, it could create trouble for society. If we try restricting knowledge creation, we will end up curtailing scientific autonomy and freedom.

The troubles associated with intellectual property rights and dangers associated with monopolies in the scientific field are yet another worry. Especially since SynBio deals with creating new entities and products, there is scope for multiple levels of patenting claims—the knowledge, the process, and the end product itself. The fear that the convergence of current IP laws with SynBio will engender cartels and monopolies, thereby increasing the commercialization of SynBio and leading to unjust scenarios, is widely discussed in this field. While some argue that patents are needed to encourage innovation and credit new inventions and discoveries, patenting underlying biological processes might be detrimental to society over time. Patents can hinder the work of more efficient competitors and inhibit or shut down research in neighbouring fieds, thereby holding back science. A good risk-benefit analysis and rethinking of the current patent system to fit a new technology like SynBio is required.

Method-related concerns such as *biosafety and biosecurity issues* arise in SynBio as well. The unpredictability and uncertainty associated with research in SynBio gives rise to many biosafety issues. The newly created synthetic entities are the first of their kind, and there is a lot of uncertainty around how they would behave and interact with the world if they escaped the specific niche designed for them. As mentioned in "Addressing biological uncertainties", "In SynBio, a circuit component well characterized in one species or strain can behave unpredictably when introduced into another due to unintended interactions with native parts" (Zhang, Tsoi, and You, 2016). In this article, the authors mention that the expression of an algal nucleotide transporter for the uptake of unnatural nucleotides caused growth inhibition in *E. coli*, which the authors attributed to the toxic effects of expressing heterologous membrane proteins. Another example would be that if newly synthesized entities were introduced into the environment, they could compete with native species, and either this interaction or the pathogenicity or toxicity of the engineered species might harm the environment.

Distribution/application-related issues are those associated with ensuring a justice-based approach in the downstream applications of research. Distributive justice focuses on the fair allocation of resources and benefits resulting from research, while procedural justice ensures that the process of distributing these resources and opportunities is fair, transparent, and inclusive. Who will have control of and access to the products of SynBio research? Would it be a monopoly yet again? How do we ensure no exploitation of human life or nature occurs during different research development stages? Once the research is completed, steps must be taken to ensure equal and efficient distribution of the benefits to all stakeholders, including the environment. The research should not widen the gap between developing countries. Extreme caution must be taken to ensure no 'helicopter research' or 'ethics dumping' occurs during the different research stages. Helicopter research occurs when researchers from high-income settings or other privileged backgrounds conduct studies in lower-income settings or on historically marginalized groups, with little or no involvement from those communities or local researchers in the research's conceptualization, design, conduct, or publication. 'Ethics dumping' occurs when similarly privileged researchers export unethical or unpalatable experiments and studies to lower-income or less-privileged settings with different ethical standards or less oversight.

The case of artemisinin is an example of a justice concern in synthetic biology. Artemisinin is a key ingredient in first-line malaria treatments recommended by the World Health Organization (WHO). It is extracted from the traditional Chinese medical herb *Artemisia annua*. According to the WHO (WHO 2018), artemisinin-based combination therapies (ACTs) provide the most effective treatment against malaria. Until 2013, natural artemisinin was sourced entirely from an estimated 100,000 small farmers in Asia and Africa, as well as wild harvesters of the crop in China. The pharmaceutical industry sources natural artemisinin from thousands of small farmers who grow *Artemisia annua*, primarily in China, Vietnam, Kenya, Tanzania, Uganda, Madagascar, and India. The average crop area per farmer in China and Africa is around 0.2 hectares. Current market demand for artemisinin is about 150–180 metric tonnes (MT). The major buyers are a handful of approved pharmaceutical companies making ACT drugs. These demands were met solely by farm-grown *Artemisia Annua* plants until the market started wavering due to climate conditions and their downstream consequences. That is when the Gates Foundation decided to step in and, supported by their funds, synthetic biologists at California-based Amyris, Inc. engineered yeast to produce artemisinic acid, a precursor to artemisinin. Pharmaceutical giant Sanofi Aventis has now scaled up commercial production to 35–60 MT of what is marketed as Semi Synthetic Artemisinin (SSA). Amyris founder Jay Keasling expressed an interest in having SSA take over full global production. In 2013, Sanofi produced 35 MT of SSA, with production rising to 50–60 MT in the coming years.

Although advocates claim synthetic biology will make anti-malarial drugs cheaper, the current production run of SSA is in fact priced at between $370–$400 per kg, significantly above the price of naturally-derived artemisinin, which sells for around $250–$270 per kg. Natural artemisinin producers further claim that it is impossible to know the true cost structure of SSA since it has received extensive philanthropic subsidies. The introduction of SSA coincided with a dramatic fall in artemisinin prices in 2013. Subsequently, in 2014, plantings of Artemesia were at only a third of previous production levels, and commercial operations were at a standstill.

Due to the production of SSA and its introduction into the market, the farmers could face a wide range of issues. It is not just the Artemisia producers who will lose a big source of their income, but also the locals who work with the downstream processes like packaging and transporting the plant. ACT's entire production and manufacturing would shift to pharma companies in the West, where malaria is less prevalent than in other parts of the world. A 2006 report from the Netherlands-based Royal Tropical Institute predicted that the SSA production could further destabilize a very young market for natural Artemisia, undermining the security of farmers just beginning to plant it for the first time. Natural producers fear the competition is unfair if SSA is marketed at a 'not-for-profit price based on large subsidies and philanthropic support from the Gates Foundation.

Apart from the impact on livelihood, another less discussed impact is the environmental impact. The lab production of most products—like semi-synthetic artemisinin or synthetic vanillin—depends on sugar, which means extensive sugarcane cultivation is required, leading to many environmental problems. For example, the increase in demand for sugar leads to an increase in the cultivation of sugarcane, which requires a lot of land and water. The surge in demand also leads to the replacement of food crops by sugarcane crops. This replacement leads to a regular monoculture of sugarcane that not only affects the biomass of the soil but also depletes nutrients in the soil, thereby affecting the ecological balance. Extensive sugarcane cultivation also contributes to rainforest deforestation and slave labour conditions.

Finally, the cultivation and agriculture of traditional plants is part of Indigenous culture and tradition which ought to be preserved, not lost in our quest for scientific discoveries and development.

Microbial cell factory

What is a microbial cell factory?

Microbial cell factories (MCFs) are gaining scientific attention for their ability to sustainably synthesize biofuels, pharmaceuticals, chemicals, and enzymes, reducing industries' environmental footprints. As a cornerstone of synthetic biology, MCFs replace resource-intensive methods with eco-friendly alternatives. In the food industry, microbial fermentation produces high-nutritional proteins and amino acids from non-edible biomass, offering sustainable solutions for animal feed, fertilizers, and alternative meat production, contributing to global food system sustainability.

What are the ethical issues?

The ethical challenges associated with microbial cell factories are multifaceted and require evaluation across the domains of biosafety and security, justice and societal impact, and philosophical considerations.

1. **Biosafety risks:** The escape of engineered microbes into the environment is a significant biosecurity concern in most microbial technologies. The potential disruption of the ecosystem, interaction with the native species, creation of new unknown species, and outcompeting of native species by engineered species are some biosafety concerns.

2. **Biosecurity concerns:** Biosecurity concerns focus on the risk of dual-use dilemmas. These arise when tools or knowledge developed for beneficial purposes are repurposed for harm, including bioterrorism or the development of antibiotic-resistant pathogens.

3. **Social concerns:** These concerns include regulatory challenges such as intellectual property claims, the creation of policies and laws regarding the use of newly designed products, and more. Patents on newly designed microbial strains or technologies can hinder further advancement and innovation and possibly hinder accessibility to the technology.

4. **Justice concerns:** Justice issues concentrate on the need to ensure equal distribution of technology's benefits and minimal to no harm to the environment, including human life. An important aspect is ensuring that the benefits of microbial cell factories are not concentrated in wealthy nations or large corporations, leaving marginalized communities at risk of exclusion from technological advances. Economic displacement of traditional industries is also a worry associated with developing new technologies.

5. **Philosophical concerns:** Concerns about 'playing God' by altering or creating new forms of life present a potential ethical hurdle for the progress of microbial cell factories. The moral status of engineered microbial strains may also face scrutiny from those who argue that using microbes solely for human benefit conflicts with ethical perspectives that recognize the intrinsic value of all life forms.

Semi-synthetic artemisinin

What is semi-synthetic artemisinin?

Artemisinin, a key malaria treatment, is traditionally extracted from the sweet wormwood plant (*Artemisia annua*). However, this method is labour-intensive and yield-dependent. Semi-synthetic artemisinin, developed using genetically engineered yeast, offers a stable supply of antimalarial drugs for high-burden regions. However, despite its medical promise, its production raises significant ethical concerns.

What are the ethical issues in semi-synthetic artemisinin production?

The ethical challenges associated with semi-synthetic artemisinin production require evaluation across the domains of biosafety and security, justice and societal impact, and philosophical concerns:

1. **Impact on farmers and local economies:** The shift to semi-synthetic artemisinin threatens the livelihoods of thousands of farmers in Asia and Africa who depend on cultivating *Artemisia annua*, often as their sole income source. This change also impacts workers in processing, packaging, and distribution, weakening local economies and raising concerns about fairness and the socioeconomic effects of technological advancements.

2. **Equity and accessibility:** Malaria remains a significant issue in the Global South, yet semi-synthetic artemisinin production is concentrated in Western nations, where malaria is less prevalent. This raises concerns about equitable benefit distribution, as high production and distribution costs could make the drug inaccessible to low-income, malaria-endemic regions.

3. **Patenting and monopoly:** Factory-based production risks centralizing control to a few corporations through patents and monopolies. This dependency could weaken the resilience of the global artemisinin supply chain, particularly during economic or political instability.

4. **Environmental justice:** Although it reduces agricultural reliance, semi-synthetic production requires significant energy and sugar inputs, raising sustainability concerns. Excessive sugarcane cultivation leads to monocropping, depletes water resources, disrupts ecosystems, and increases the carbon footprint.

5. **Loss of traditional knowledge systems:** The complete transition to lab-based production risks eroding traditional farming practices and the ecological knowledge embedded in them. Balancing technological innovation with preserving traditional systems is essential for ecosystem protection and cultural heritage.

Genetic modification

What is genetic modification?

Genetic modification involves altering an organism's genetic make-up using techniques such as CRISPR-Cas9 for precise gene editing, gene insertion to enhance desirable traits, or synthetic biology to create new genetic functions (see also the Health Care Ethics chapter). It can serve various purposes, including curing diseases, preventing the inheritance of specific genes, or enhancing physical, cognitive, and behavioural traits. In humans, genetic modification might improve intelligence, strength, or longevity; while in agriculture, it could boost crop yields or pest resistance. Despite its potential benefits, genetic modification raises significant ethical concerns.

What are the ethical issues related to genetic modification?

1. **Global access and inequity:** Genetic enhancement could widen existing social inequalities between those who can afford modifications and those who cannot. This could lead to the emergence of 'genetic elites' with advantages in education, employment, and social status, deepening global divides and raising concerns about fairness and fear of stratification in the society.

2. **Risk of eugenics:** Genetic enhancement might revive eugenic ideologies, promoting the idea of designing 'better' humans. The practice may revive eugenic ideologies by stigmatizing undesirable traits, reinforcing discrimination, and leading to the marginalization of individuals with disabilities or differences.

3. **Threat to individual autonomy:** The normalization of genetic modification could pressure individuals to participate in it for social or professional benefits, undermining their personal choices and autonomy.

4. **Germline genetic enhancement and unintended consequences:** Heritable genetic modifications raise concerns about consent, unforeseen health risks, and disruptions to biological systems, affecting future generations. Modifications could also have unforeseen impacts on human health, including increased susceptibility to diseases or disruptions to complex biological systems.

5. **Slippery slope to non-essential enhancements:** Approving genetic modifications in order to improve health could slowly blur the line between necessary and optional modifications. This could eventually lead to a slippery slope of non-essential enhancements, like improving skin texture, changing eye colour, and increasing physical strength or mental capability.
6. **Environmental and ecological implications:** Genetic modifications in agriculture to improve the physical traits of livestock crops can have a detrimental effect on the ecosystem, because they might lead to unforeseen ecological imbalances which eradicate natural populations in the wild

SynBio and non-dualism

All concerns in SynBio challenge distinctions such as life vs machine, natural vs unnatural, and life vs non-life and question the role of humans in creation, the boundaries in terms of trying to protect nature, and the moral status of the newly created entities. In current ethical literature on synthetic biology, the distinctions between life and non-life, biology and technology, and natural and unnatural carry normative weight. The fact that synthetic biology challenges these distinctions is considered ethically relevant. At the same time, the common factor among many ethical concerns surrounding SynBio is that they begin from a dualistic assumption. For example, they assume that the moral status of these created entities hinges on the answer to the question of whether they pertain to the domains of 'life' or 'non-life'. The fact that humans—or, more specifically, synthetic biologists—are now at the threshold of constructing living beings or parts of living beings from scratch raises questions about their authority to create life from scratch. One of the biggest worries in ethical literature is the scientists' role in creating entities that have never existed before. Concerns start with the risk of scientists 'playing God' by creating life, the moral status of these newly created entities, and the essence of human relationships with nature.

Let us try considering these concerns through a non-dualistic approach or framework to find a possible way to address them. This section will use ancient Indian philosophy to situate and address these philosophical concerns through the Indian philosophical framework. Ancient Indian Hindu philosophy is an example of biocentrism in which, though a human being is thought to be endowed with a consciousness that exceeds the consciousness of other species, they are not considered superior to other species. Hinduism takes a holistic approach to life and nature which considers each human being an integral part of an organic whole, and the natural forces are considered sacred. In the spiritual, metaphysical view of Hinduism, human life—like every other life on earth—forms part of the web of existence. Together with material elements, human and non-human species are indissolubly linked to an organic whole, thereby remaining non-dualistic in their approach.

To better understand any philosophical system or theory, including Hinduism, it is essential to understand its origin. India was originally referred to as 'Bharath' in Sanskrit, where *Bha* refers to light, knowledge, and effulgence, and *Rath* means 'in search of'. Bharath essentially meant 'in search of light'. This was a geographical identity used to denote the land with a conglomeration of different kingdoms, big and small, all bound by a common culture—the culture of experiencing divinity in all aspects of life. The people of this land lived a particular way of life in sync with nature called the 'Sanatana Dharma', which we now know as 'Hinduism'. The roots of the phrase 'Sanatana Dharma' can be traced back to ancient Sanskrit literature as a kind of cosmic order in which *Sanatana* denotes 'that which is without beginning or end' or 'everlasting', and *Dharma*—coming from *dhri*—means 'to hold together or sustain'. *Dharma* is often interpreted as meaning 'natural law'. 'Sanatana Dharma' can thus be understood as 'eternal duties' or natural way to live'.

In Sanatana Dharma, Atman, and Brahman are two metaphysical concepts, where Atman is the individual self and Brahman is the ultimate reality, the supreme being. While Brahman is the divine essence of the universe, Atman is the essence that lives in all: humans, non-humans, and nature. According to the philosophy of Sanatana Dharma, the ultimate reality of everything in the universe is the Brahman which is attribute-less (nirgun), formless (nirakar), infinite (anant), and omnipresent (sarvabyapi). 'Sarvam khalvidam brahma' (everything that has existence is Brahman; Chhandogyopanishad) (Awasthi, 2021). The concept of Brahman in Santana Dharma is very similar to the concept of Tao in Taoism, both conveying something that rational thoughts or words cannot convey (Brahman, the Tao, and the Ground of Being, 2016).

Sanatana Dharma sees only one reality or being, Brahman, which all different living and non-living forms are born from and assimilate back into after death. This is similar to the thought that "everything that exists is nature" (Ducarme and Couvet, 2020). In other words, in this culture, 'God' is not a supreme being among lesser, subordinate beings; instead, all beings are a manifestation of the one reality or being called Brahman. In this culture, God (Brahman) is omnipresent and resides in everyone and everything, including all living and non-living things, thereby blurring the difference between living and non-living.

Humans and nature or non-human forms are seen as separate entities in dualistic frameworks. Sanatana Dharma, one of the non-dualistic frameworks, views nature as inclusive of all forms in this universe—which are seen as contiguous components of a hierarchical order of beings, related to each other through a network of functional and natural relationships based on their location in this order. The order is maintained by a universal natural law, sometimes called *Rta* (pronounced rita). Meera Baindur, in her work *Nature as Non-Terrestrial*, explains the diversity of beings and their relationships to each other and details how the inner being or consciousness is viewed as one in all forms in this universe (Baindur, 2009).

While thinking about the role of human beings in this world, as per Sanatana Dharma—there is no inherent superiority of any species—humans are neither co-creators with 'God' nor stewards of nature; superiority, if any, is only in living, and upholding Dharma. It is proposed that while humans have the advantage of equipment and methods such as Jnana, karma, bhakti, and raja yoga compared to other forms of life, it is the quality of 'atma-vichara' (self-contemplation) that might be unique to human beings.

सर्वभूतस्थमात्मानं सर्वभूतानि चात्मनि ।
ईक्षते योगयुक्तात्मा सर्वत्र समदर्शन: ॥ 29॥

BG 6.29: The true yogis, uniting their consciousness with God, see with an equal eye all living beings in God and God in all living beings. (Mukundananda, 2014)

As per Sanatana Dharma, humans have a responsibility towards nature and all forms of life, not as co-creators but as people who are a part of the web of nature. This is often enacted through various forms of 'non-violence'—a general term attributed to India and Hinduism. However, it is essential to point out that while Indian Hindu philosophies stress the importance of non-violence toward all creatures, it does not only mean non-violence through action but also non-violence in thought, word, and deed. This is seen as the highest of all forms of righteousness or *dharma*.

Based on the principles of Sanatana Dharma, every entity—irrespective of its origin—possesses moral status and value since all forms and entities in this universe (living and non-living) are parts of nature and comprise of the same five elements known as *panchabhutas*, or *panchamahabhutas*, in Sanskrit. They form the basic building blocks of the universe; every person, animal, plant, and thing is composed of various combinations of the *panchabhutas*, thereby removing any difference between the living and non-living as we have been viewing them. This view also stems from a belief in the concept of reincarnation. Hindu teachings suggest that the human soul can reincarnate in any form, including the forms of lesser and simple living organisms, as well as more complex forms this gives rise to the *dharma* of treating everyone and everything with respect and reverence.

Science and culture or religion are part of the same system. As in most non-Western philosophies, the non-dualism of science and culture or traditions is another feature of Sanatana Dharma. In the book *Research is Ceremony*, author Shawn Wilson quotes the Mayan scholar Carlos Cordero (1995): "The difference within the Western knowledge system is that there is a separation of areas called science from those called art and religion. On the other hand, the [Indigenous] knowledge base integrates those areas of knowledge so that science is both religious and aesthetic" (p. 55). While in most parts of the West, knowledge is approached using intellect, most non-Western and Indigenous cultures approach knowledge through senses and intuition. Sanatana Dharma has a holistic understanding of everything in the universe, where the universe and every small entity are understood as a whole and not studied or viewed as separate parts.

As per Sanatana Dharma, the universe came into being by the wish of Lord Vishnu, and Lord Vishnu maintains the entire universe and its cosmic balance. While this might be a story or a superstition for many, the beautiful part is the intricate meaning behind the name 'Vishnu'. In Sanskrit, *Viṣṇu* includes the root *viś*, meaning 'to settle, to enter', or also (in the Rigveda) 'to pervade'. While the Sanskrit term cannot be translated completely faithfully into English, something close would be 'all-pervasive'. Someone who is everything and is found in everything. The name indicates the whole universe in one, irrespective of our forms. This is one of the reasons why Sanatana Dharma, like other Eastern and Indigenous concepts, emphasizes the importance of transcending the material world and finding one's true divine inner nature and one's place in the universe.

Hindu philosophy is one example of the many non-dualistic philosophies around the world. Being open to embracing viewpoints from these philosophies allows us to widen our perspectives and view ethical concerns in a different light. In synthetic biology, the philosophical and anthropological concerns that often dominate scholarly engagement are largely based on dualisms that separate human life from nature. If such dualisms and the questions they raise can be addressed through taking up a non-dualistic stance, we may be able to refocus our efforts on other social concerns that require our urgent attention.

Conclusion

Ethical considerations in synthetic biology (SynBio) extend beyond biosafety and biosecurity concerns to encompass issues of justice and politics, necessitating an approach that accounts for both theoretical frameworks and the practical realities of laboratory research. Given SynBio's interdisciplinary nature and dual-use potential, ensuring its overall positive impact requires a stage-wise and research area-specific ethical analysis, rather than treating it as a monolithic technology. Ethical assessments should be integrated at distinct phases of research—knowledge generation, methodological development, and application—while also prioritizing environmental and livelihood justice to address broader societal implications. Establishing ethical awareness early in researchers' careers can foster a long-term commitment to responsible research practices, influencing both individual projects and institutional policies. Although political and corporate interests often drive technological development, fostering public engagement and ethical discourse remains imperative. Furthermore, current ethical discussions on SynBio frequently rely on dualistic frameworks, such as nature versus machine or life versus nonlife, which can lead to conceptual deadlocks. Integrating non-dualistic perspectives, particularly from non-Western philosophies, can provide deeper insights and contribute to a more holistic and context-sensitive ethical approach to SynBio.

Bibliography

Aravind Paleri, Varsha, and Kristien Hens. 2023. "Beyond the Organism versus Machine Dichotomy: A Review of Ethical Concerns in Synthetic Biology". *ACS Synthetic Biology* 13 (1): 3–14. https://doi.org/10.1021/acssynbio.3c00456

Baindur, Meera. 2015. *Nature in Indian Philosophy and Cultural Traditions*. Vol. XII. Sophia Studies in Cross-Cultural Philosophy of Traditions and Cultures. New Delhi: Springer India. https://doi.org/10.1007/978-81-322-2358-0

Baindur, Meera. 2009. "Nature as Non-Terrestrial: Sacred Natural Landscapes and Place in Indian Vedic and Purāṇic Thought". *Environmental Philosophy* 6 (2): 43–58. https://doi.org/10.5840/envirophil20096213

Basso, L. C., Thiago Olitta Basso, and Saul Nitsche Rocha. 2011. "Ethanol production in Brazil: The industrial process and its impact on yeast fermentation". In Biofuel Production—Recent Developments and Prospects, edited by Marco Aurelio dos Santos Bernardes, pp. 85–100. Rijeka: IntechOpen. https://doi.org/10.5772/17047

Benner, Steven A., and A. Michael Sismour. 2005. "Synthetic Biology". *Nature Reviews Genetics* 6 (7): 533–43. https://doi.org/10.1038/nrg1637

"Case Study: Artemisinin and Synthetic Biology | ETC Group", 2 July 2014. https://www.etcgroup.org/content/case-study-artemisinin

Douglas, Thomas, and Julian Savulescu. 2010. "Synthetic Biology and the Ethics of Knowledge". *Journal of Medical Ethics* 36 (11): 687–93. https://doi.org/10.1136/jme.2010.038232

Ducarme, Frédéric, and Denis Couvet. 2020. "What Does 'Nature' Mean?". *Palgrave Communications* 6 (1): 1–8. https://doi.org/10.1057/s41599-020-0390-y

European Academies Science Advisory Council (EASAC). *Synthetic Biology: An Introduction*. January 2011. https://www.interacademies.org/sites/default/files/publication/synthetic_biology_an_introduction_feb_2011.pdf

Gibbs, W. Wayt. 2004. "Synthetic Life". *Scientific American* 290 (5): 74–81.

Hudson Robotics. n.d. "Introduction to Synthetic Biology: Exploring the Basics and Applications". Accessed 11 July 2024. https://hudsonrobotics.com/introduction-to-synthetic-biology-exploring-the-basics-and-applications/

Kitney, Richard, and Paul Freemont. 2012. "Synthetic Biology - the State of Play". *FEBS Letters* 586 (15): 2029–36. https://doi.org/10.1016/j.febslet.2012.06.002

Lakoff, George, and Mark Johnson. 2003. *Metaphors We Live By*. Chicago: University of Chicago Press. https://press.uchicago.edu/ucp/books/book/chicago/M/bo3637992.html

"Nature Addresses Helicopter Research and Ethics Dumping". *Nature*, 30 May 2022. https://doi.org/10.1038/d41586-022-01423-6

Okafor, Justus Onyebuchi, and Osim Stella. 2018. "Hinduism and Ecology: Its Relevance and Importance". *FAHSANU Journal* 1 (1). https://philarchive.org/rec/OKAHAE

Ouaray, Zahra, Steven A. Benner, Millie M. Georgiadis, and Nigel G. J. Richards. 2020. "Building Better Polymerases: Engineering the Replication of Expanded Genetic Alphabets". *The Journal of Biological Chemistry* 295 (50): 17046–59. https://doi.org/10.1074/jbc.REV120.013745

"Scientific Committee on Health and Environmental Risks (SCHER) - European Commission". Accessed 11 July 2024. https://health.ec.europa.eu/scientific-committees/former-scientific-committees/scientific-committee-health-and-environmental-risks-scher_en

Szczebara, F. M., Cathy Chandelier, Coralie Villeret, Amélie Masurel, Stéphane Bourot, Catherine Duport, Sophie Blanchard, Agnès Groisillier, Eric Testet, Patricia Costaglioli, Gilles Cauet, Eric Degryse, David Balbuena, Jacques Winter, Tilman Achstetter, Roberto Spagnoli, Denis Pompon, and Bruno Dumas. 2003. "Total biosynthesis of hydrocortisone from a simple carbon source in yeast". *Nature Biotechnology* 21 (2): 143–149. https://doi.org/10.1038/nbt775

"When Synthetic Biology Meets Medicine | Life Medicine | Oxford Academic". Accessed 11 July 2024. https://academic.oup.com/lifemedi/article/3/1/lnae010/7623268

Widmaier, D. M., Danielle Tullman-Ercek, Ethan A Mirsky, Rena Hill, Sridhar Govindarajan, Jeremy Minshull, and Christopher A. Voigt. 2009. "Engineering the Salmonella type III secretion system to export spider silk monomers". *Molecular Systems Biology* 5 (1): 309. https://doi.org/10.1038/msb.2009.62

Wilson, Shawn. 2008. *Research Is Ceremony: Indigenous Research Methods*. Nova Scotia: Fernwood Publishing.

World Health Organization (WHO). "Artemisinin resistance and artemisinin-based combination therapy efficacy: Status Report". December 2018. https://www.who.int/docs/default-source/documents/publications/gmp/who-cds-gmp-2018-26-eng.pdf

Zhang, Carolyn, Ryan Tsoi, and Lingchong You. 2016. "Addressing Biological Uncertainties in Engineering Gene Circuits". *Integrative Biology* 8 (4): 456–64. https://doi.org/10.1039/c5ib00275c

Zhou, Y., et al. 2023. "Improved production of Taxol® precursors in Saccharomyces cerevisiae using a genome-scale metabolic model-guided approach". *Microbial Cell Factories* 22 (1): 221. https://doi.org/10.1186/s12934-023-02251-7

8. Literary Bioethics

Introduction

In the novel *Never Let Me Go* (2005) by Nobel Prize winner Kazuo Ishiguro, cloned human beings who are genetically identical to their non-cloned counterparts are used by the government as organ donors. These clones feel, act, and think just as regular human beings but do not have equal rights in their society because they do not meet the exact requirements for 'humanness'. The (bio)ethical questions that this example raises show how fiction and bioethics interact multi-directionally: bioethics is not only a source of inspiration for writers, but fiction can also be useful for (bio)ethical practice. Fiction and cultural imagination, in general, are also par excellence places where possible answers to bio(ethical) dilemmas can be explored.

For a long time, the term 'moral imagination' was considered an 'oxymoron' (Johnson, 2016). Moral philosophy has traditionally—and increasingly so since modernity—placed rationality, rules, and principles at the heart of morality. Moral judgement and moral action are usually regarded as (the outcome of) a rational process that must not be tainted by the volatile and possibly illusory nature of imagination. However, several authors have challenged this view by stressing instead the need for imagination in morality. Most have done so by conceiving moral imagination in moral reasoning as (a) empathy, (b) imaginative perception, or (c) a phase of deliberation.

The now quite popular idea of (a) empathy as an important moral tool stretches back to Adam Smith's (1790) hypothesis that we need imagination to understand each other's problems. In this vein, Martha Nussbaum described moral imagination as "the ability to think what it might be like to be in the shoes of a person different from oneself" (1997, pp. 19–11). Following Iris Murdoch's claim that "clear vision is a result of moral imagination and moral effort" (2001, pp. 35–36), other philosophers have presented accounts of moral imagination as (b) an imaginative apprehension of reality through which we get a better grip on other persons and particular situations and events (see Nussbaum, 1985; Diamond, 1991; Chappell, 2017; Ratajczyk, 2024). From a very different pragmatist perspective, John Dewey (1922) regarded imagination as (c) a crucial phase of deliberation he called 'dramatic rehearsal'. Scholars following Dewey regard such dramatic rehearsal as an experimental phase of action during which we imaginatively try out different actions to discover their possible consequences, appropriateness, reception, and feasibility (see also Fesmire, 2003).

Mark Johnson (1993) went beyond illustrating a place for imagination and moral reasoning. He argued that all of our fundamental moral concepts (such as duty, action, utility, virtue, etc.) have a metaphorical basis and thus rest on imaginative language use. Therefore, he claims that moral reasoning is imaginative and moral philosophy should therefore switch its usual focus on rule-following and principle application to uses of imagination (see also Chappell, 2017 for a similar anti-theoretical take on moral philosophy). Coeckelbergh (2007) has mitigated such claims and has observed complementary roles for imagination and principle application in moral reasoning.

Variations on or combinations of the aforementioned views have found their way into applied ethics, such as business ethics (Werhane, 1999), peacebuilding (Lederarch, 2005), and medical ethics (Coeckelbergh and Mesman, 2007). But also in bioethics, often dealing with cutting-edge or even speculative technologies, imagination can play an important role.

In her book *Literary Bioethics* (2020), Maren Tova Linett gives several arguments for the use of fiction for bioethics:

- Fiction (literature, film, etc.) quickly picks up scientific findings and helps us reflect on science's storytelling.
- Fiction helps us imagine questions of life (philosophical questions, ethical dilemmas) and can indirectly comment on them.
- When we read and analyze fiction, it helps us to shed light on assumed knowledge. It helps us make certain biases (anthropocentrism, for example) explicit.
- Fiction can help us examine a theme or issue from a different perspective because stories can take us out of our comfort zone or habitus.
- Fiction can help us to (re)value different kinds of lives. Literature has often reproduced restrictive and stigmatizing norms about who or what counts as a human being or living being, but it can also question existing ideas about this.
- Imagination and specificity in literature helps with countering abstractions of humans and other living beings different from us. Suzanne Keen (2007) discusses the potential for fictional works to increase empathy. At the same time, she argues that it is also important to be critical of the so-called 'empathy-altruism hypothesis', which proposes an overly direct link between reading fiction and our concrete altruistic behaviour.

Literary form and genres

"It matters what stories we tell to tell stories with; it matters what concepts we think to think other concepts with", as the philosopher Donna Haraway famously put it (2019,

p. 10). Showing 'how we world our world' and how its 'reality' might be reinvented is perhaps a specialty of science-fiction author Ursula Le Guin (author of *A Wizard of Earthsea*, 1968; *The Left Hand of Darkness*, 1969; *Always Coming Home*, 1985). In her essay *The Carrier Bag Theory of Fiction* (1988), she redefines the story as cultural technology in a feminist way. Contrary to dominant narratives shaped around the quest of a hero, the killing of an enemy, etc., stories should be shaped like a bag—the first tool of humanity—to offer room to hold things that bear meanings and enable relationships. As Le Guin defines it, a story's purpose is neither resolution nor stasis but a continuing process. To Le Guin, science fiction (like all serious fiction) is a way to describe what is going on, what people do and feel, and how people relate to everything.

The Carrier Bag Theory of Fiction was originally published in 1988. However, it recently experienced a revival due to Donna Haraway's appreciation of it in her book *Staying with the Trouble* (2016) and a subsequent reissue of the essay in 2019. It is not surprising that many of the main characters in Le Guin's stories are anthropologists. These anthropologists attend to multiple ways of living and telling. They use culture as a tool of analysis and see culture as a tool to shape worlds.

Parallel to this, non-Western thinkers and storytellers can help us shed light on biases engrained in Western culture and language. In her bestselling book *Braiding Sweetgrass* (2013), Robin Wall Kimmerer—who introduces herself as a mother, botanist, and member of the Potawatomi community—weaves together Indigenous wisdom, scientific knowledge, and the teachings of plants. She occupies a unique position on the intersection of science and indigenous knowledge that allows her the authority to dismantle the exclusivist thinking of Western science. In the chapter "Learning the Grammar of Animacy", she shows that the English language consists mostly of nouns. Words such as 'bay', 'Saturday', 'hill', and 'beach' are all nouns for humans to refer to the lifeless world. In contrast, in Ojibwe or Potawatomi—two Indigenous languages she refers to—there are verbs to describe these same phenomena: e.g. *wiikwegamaa* means 'to be a bay', and there are other verbs meaning 'to be Saturday', 'to be a hill', and 'to be a beach'. Instead of considering this a mistake, we could learn from it. In Potawatomi and most Indigenous languages, the same words are used to refer to the living world as to family. When Kimmerer goes into the woods with her students, she tries to be bilingual: to teach her students both the scientific language and the grammar of animacy, to look at the woods as a communion of subjects, not a collection of objects. She teaches her students by taking them out into the fields and by being mindful of how she speaks of the organisms they study as kin.

Genre and different media

When standard use of our scientific language does not suffice to describe these phenomena, fiction in all its forms is there to help us to imagine and think. Fiction (novels, but also film, video games, graphic novels, poetry, etc.) allows us to capture different points of view, events, and objects in a story. According to some literary critics, poetry has more freedom in its use of language, which makes it perfectly fit to practice a grammar of animacy, as Kimmerer described, or to bring together subjects that live on different scales, to switch between subjects and objects. Poetic language makes use of the material of language and has a strong affective component. More than a description of events, it allows for an expression of the emotive movement.

Science fiction

Science fiction, the counterpart of realism in literature, has existed for a long time, but increasingly, its central speculative elements are based in observations about the climate crisis. This literature increasingly thematizes ecology, global warming, petroculture, and pollution. The 'weirdness' of these books is alienating, helping to problematize the biopolitical status quo. We already mentioned Ursula Le Guin's work. Another author who makes use of the speculative element in science fiction to think through our coexistence with fungi is Jeff Vandermeer in his *Southern Reach* trilogy, of which the first book—*Annihilation*—was published in 2014.

Although contemporary science fiction is often dystopic, new aesthetic subgenres continue to emerge. Solarpunk, for example, is a scientific subgenre that emerged online around 2008, which visualizes collectivist ecological utopias where nature and technology grow in harmony.

Ecofiction

The subgenre of ecofiction perfectly illustrates thinking at the intersection of environment, literature, and ethics, as well as the way fiction can stimulate moral reflection. Nature has always been a popular theme in fiction but often served as a backdrop to or metaphor for the development of human characters. Romantic poets such as William Blake (1757–1827) and William Wordsworth (1770–1850) wrote in a nostalgic tone about nature as a pristine idyll and as something pure and innocent. On the other hand, nature was presented as a landscape against which the enlightened individual could develop himself, as happens in the autobiographical *Walden* (1854) by the American writer Henry David Thoreau. Even in more recent literature, such as John Krakauer's *Into the Wild* (1996)—about the true story of Christopher McCandless, who went into the remote parts of Alaska—taming and finding one's way in nature is the common thread within the story.

But in recent decades, in light of growing awareness of climate change and the consequent climate crisis, fiction about nature is changing. Nature is no longer just the background of a story about humans: it takes on an active role in the literary text, becomes an actor. Nature is no longer presented merely as an unknowable and omnipotent wilderness in which the human protagonist is rendered powerless. Instead, through the thematization of pollution, desertification, deforestation, and flooding, the vulnerability of nature is also emphasized. Many recent literary works thematize the paradox between the vulnerability of nature and its destructive power in relation to humans.

On the one hand, the climate crisis recurs universally in literature, but on the other hand, these themes are also *situated*. In the Dutch-speaking world, for example, we see many recent artworks about a possible rise in the sea level and its disastrous consequences for those living below sea levels. Examples include the novel *Zee Nu* (2022) by Eva Meijer, and the political foundation and art project 'Embassy of the North Sea'.[1] The latter was founded to listen to, speak with, and negotiate on behalf of the sea and marine life. In *The Man with the Compound Eyes* (2011) by Taiwanese author Wu Ming-Yi, the lead role is played by the Pacific Ocean's plastic soup that permanently alters Taiwan's coastline and weather. The plastic soup forms a mountain and, in this way, almost becomes part of the landscape. The novel narrates how the soup clashes but also becomes integrated into the lives of the indigenous people on the islands in the ocean, and how they affect each other.

(Bio)ethical questions in fiction

(Bio)ethical questions explored in fiction can be broadly divided into three categories: human-human relations, human and other-than-human relations, and human-planet relations. We will explain each of these below and give concrete examples of literary and artworks and the questions they raise.

Human-human relations

In *Literary Bioethics* (2020), Maren Linett discusses several literary works that raise bioethical questions about inter-human relationships. These include the absence of old people in Aldous Huxley's *Brave New World* (1932) and what that says about the perception of old age in our society, or the contested value of intellectually disabled people in Flannery O'Connor's *The Violent Bear It Away* (1960).

These examples show how developments and attitudes around health, health care, and medicine are reflected in cultural expressions in society. In addition, narratives of lived experience are an important source for knowledge formation on these topics. Therefore, medical and clinical perspectives alone cannot answer contemporary questions around themes of health, illness, disability, patient experience, sustainable care, life, and death. Thus, medical and clinical perspectives must be equally complemented by philosophical, historical, anthropological, sociological, narratological, and cultural perspectives. Much research on this is being done in the academic disciplines of 'medical humanities' and 'health humanities'.

---

1 https://www.embassyofthenorthsea.com/

There are several concrete ways in which works of fiction can be used to improve interpersonal relationships. A first example is the creation of more insight and understanding into the lifeworld of others, for example through autobiographical stories. A second example is actively questioning and challenging existing stereotypes. Hilde Lindemann, who specializes in narrative ethics, writes in *Damaged Identities, Narrative Repair* (2001) about, for instance, the potential of counter-stories—which can become a strategy of resistance for oppressed groups that allows their stigmatized identity to be narratively 'repaired.'

Human and non-human relations

Another theme that has been explored in fiction is how humans interact with other species. Although the definition of what 'humans' are is forever under debate, 'human' is often a category that allows for certain rights other-than-human animals do not have self-evident access to. Animal studies ask us to think about what qualities grant rights and, more generally, moral value. They ask us to stop before we dismiss a living being, human or other-than-human, from the sphere of moral consideration and question the basis for such a dismissal. Literary thought experiments have often explored these themes and carry a long tradition.

One example of an old literary work is an epistle written by the Brethren of Purity in tenth-century Iraq: *The Case of the Animals versus Man before the King of the Jinn* (Ikhwān al-Ṣafāʾ, 2009). Seventy men are shipwrecked on an island that is inhabited by animals who have fled from the descendants of Adam to avoid exploitation. When the men attempt to subjugate the animals, the animals demand justice, and a trial is convened before the king of the Jinn. The argument of the humans is that mankind is superior to animals. The animals—represented by the bee, the jackal, the parrot, the frog, the cricket, and an unspecified cattle ambassador, probably a mule—refer to the Quran to suggest that mankind does not have exclusive rights over creation, that bees have a skillset superior to that of humans, that mankind is confused, given the existence of world religions. Although the trial concludes with a verdict on mankind's behaviour, the epistle remains an interesting example of an early weighing of moral judgement considering non-human lives.

Another example is *The Lives of Animals* (1999), a meta-fictional text written by Elizabeth Costello: the alter-ego of the Nobel-prize-winning author J. M. Coetzee. It consists of two chapters, 'The Philosophers and the Animals' and 'The Poets and the Animals'. The author—whose own views are made ambiguous through his playing with form and content and use of an alter ego—discusses the foundations of morality and proposes an ethic of 'sympathy', that leaves the initiative for the sympathy up to the subject: "there are people who have the capacity to imagine themselves as someone else, there are people who do not have such capacity" (1999, p. 41).

Gunda, a recent (2020) documentary in black and white made by Victor Kossakovsky, is an intimate account without words of the life of a mother pig, a flock of chickens, and a herd of cows. The documentary breaks expectations for the narrative of traditional film and instead provides space for the animals' daily actions and movements, allowing viewers to build an intimate relationship with them and pointing to what these animals have in common with human animals: the love of a mother for her child. The film indirectly invites its viewers to reflect on themes such as agro-industry, meat consumption, and related themes.

Recent scientific insights on the lives and intelligence of other-than-human animals and beings, together with recent posthumanist currents in philosophy, have challenged human exceptionalism. Contemporary fiction and popular science publications have expanded our knowledge of other forms of consciousness, intelligence, and life. Zooming in on the life of another living being through fiction can help to dismantle humans' exceptional status and expand our sympathy towards other-than-human animals or beings.

Human-planet relations

Our understanding of the current geological era as the 'Anthropocene'—a term which recognizes humanity's growing influence on the planet—also affects fiction. A useful concept here is Timothy Morton's idea of the hyperobject (Morton, 2013). This is an 'object' that is so spread across time and space that it becomes elusive but is still named as an object, such as climate change, capitalism, or the internet. We can only interact with a part of it. Fiction can make planetary scale or climate issues imaginable and comprehensible by zooming out beyond the individual perspective, both in terms of time and space.

Someone who makes this call for a different approach to time and space in literature explicitly is the Indian writer Amitav Ghosh. He writes both fiction and nonfiction (*Gun Island*, 2019; *The Nutmeg's Curse*, 2021) and argues in *The Great Derangement* (2016) that contemporary, realist modernist fiction can also mask climate issues through too much focus on individual emotions and mundane details. He argues for the need to write literature on a planetary scale and to clarify how the individual or local scale relates to the global and even planetary scale. As such, not only should the different scales of life on Earth be highlighted, but the interaction between them should also be stressed in order to highlight the implication of individual humans in the larger whole.

On the other hand, in order to imagine planetary life, it may actually be important to scale down and zoom in to the microscopic or microbial scale. Microbes are the origin of all life. Without bacteria, viruses, and other single-celled or acellular organisms, life as we know it would cease to exist, yet they escape daily human observation. In *Life, Re-Scaled* (2022), Liliane Campos and Pierre-Louis Patoine write about the need to zoom in to the invisible scales of microbial or neurochemical phenomena, which can very well be examined through an aesthetic that emphasizes defamiliarization.

Ultimately, fiction might also zoom out from a Western, short-term time scale to 'deep time' or a geologic time perspective to make this phenomenon comprehensible.

Literary scholar Mark McGurl describes this as 'planetary indifference' (2012): the planet can do without humans and might even be better off without us. At the same time, this also requires critical thinking about the implications of our use of artificial materials, which will continue to exist long after we are gone. The poetry of Flemish poet Dominique De Groen[2] (2019) thematizes the clash between raw, Earth-derived materials and artificial materials. Her poetic language makes ecocritical ideas palpable by mixing them with the materiality of contemporary life.

> **Exercise:** Read the poem by Dominique De Groen below and reflect on:
>
> What metaphors and images are being used to capture the theme—and what is the effect of this?
>
> - How does the poem interact with contemporary scientific discourses?
> - How does the medium help to construct a story and stimulate the imagination?
> - How does the subject influence the aesthetic? How does the poem work around the question of scale?
> - What philosophical/ethical questions does the poem raise?
> - Does the poem introduce new perspectives on the value of life/coexistence?

XI. ICE

The plains stretch out endlessly
white under the burning sun:

an expansive landscape of dazzling plastic
bathing in the green-tinted twilight
of a nuclear winter.

On the black, Arctic ice
we hunt for algae

gray doubles
of the first people
we chew on a bitter, grinning plant.

A bleached, fragile coral reef
buried under waves of sand

---

2 Dominique De Groen (1991) is a Belgian poet and writer. She has published the poetry collections *Shop Girl* (2017), *Sticky Drama* (2019), *Offerlam* (2020), and *SLANGEN* (2022). Her work has been awarded with the Frans Vogel Poetry Prize 2019, the audience prize of the Fintro Prize for Literature 2021, and the Jan Campert Poetry Prize 2022. In her poetry, fiction and essays she engages with themes like ecology, (anti-)capitalism, globalisation, and internet and pop culture.

offers our soft bodies protection:

the dead reef
is still porous
absorbing the miniscule, poison-green crystals
of toxic radiation.

The surviving plants
are black and sticky
like oil.

They grin:
the hegemony
of the sun's capitalism
the economy of photosynthesis
is coming to an end.

From slumbering bacteria
deep in the black ice
they steal their sustenance.

With voices of slime
they whisper of the revolution

a new era

a wet, dark
subterranean sun.

My body overflows
with molten creatures
waiting to congeal
and be reborn.

The ur-slime
that slowly washes over me
is black and wet

reeks of the guts
of an iceberg.

The spirits of the old animals
are dazzling in the nuclear night.

Conclusion

We examined how fiction, particularly speculative and literary fiction, can enrich bioethical reflection by engaging the moral imagination. This discussion began with *Never Let Me Go* (2005) by Kazuo Ishiguro, where cloned individuals—emotionally and cognitively indistinguishable from humans—are denied full moral status, prompting ethical questions about personhood and human dignity. We drew on philosophers such as Martha Nussbaum, Iris Murdoch, and John Dewey, to highlight how imagination—whether in the form of empathy, perception, or deliberation—is not opposed to moral reasoning but central to it. Fiction, we argue, can illuminate implicit biases, invite new perspectives, and explore ethical dilemmas in ways that abstract principles often cannot. Through works by authors such as Ursula Le Guin, Robin Wall Kimmerer, and Maren Linett, we demonstrated how literature can challenge dominant narratives, expand notions of moral considerability, and reframe relationships between humans, non-humans, and the planet.

Bibliography

Babbitt, S. 1996. *Impossible Dreams: Rationality, Integrity and Moral Imagination*. Boulder: Westview Press.

Campos, Liliane, and Pierre-Louis Patoine, eds. 2022. *Life, Re-Scaled: The Biological Imagination in Twenty-First-Century Literature and Performance*. 1st edition. Cambridge: Open Book Publishers. https://doi.org/10.11647/OBP.0303

Chappell, S. G. 2017. *Knowing What To Do: Imagination, Virtue and Platonism in Ethics*. Oxford: Oxford University Press.

Coeckelbergh, M. 2007. *Imagination and Principles. An Essay on the Role of Imagination in Moral Reasoning*. Basingstoke: Palgrave Macmillan.

Coeckelbergh, Mark, and Jessica Mesman. 2007. "With Hope and Imagination: Imaginative Moral Decision-Making in Neonatal Intensive Care Units". *Ethical Theory and Moral Practice* 10 (1): 3–21.

Coetzee, J. M. 1999. *The Lives of Animals*. Princeton: Princeton University Press.

De Groen, D. 2019. *Sticky Drama*. Aalst: Het balanseer.

Dewey, J. 2002. *Human Nature and Conduct*. New York: Dover Publications.

Diamond, C. 1991. "Missing the Adventure". In *The Realistic Spirit: Wittgenstein, Philosophy, and the Mind*, edited by C. Diamond, pp. 309–318. Cambridge: MIT Press.

Fesmire, S. 2003. *John Dewey & Moral Imagination: Pragmatism in Ethics*. Bloomington: Indiana University Press.

Ghosh, A. 2016. *The Great Derangement: Climate Change and the Unthinkable*. Chicago: University of Chicago Press.

Ikhwān al-Ṣafāʾ (Brethren of Purity). 2009. *The Case of the Animals versus Man Before the King of the Jinn: An Arabic Critical Edition and English Translation of Epistle 22*. Edited and translated by Lenn E. Goodman and Richard McGregor. Oxford: Oxford University Press in association with the Institute of Ismaili Studies.

Ishiguro, Kazuo. 2005. *Never Let Me Go*. London: Faber and Faber.

Johnson, M. 2016. "Moral Imagination". In *The Routledge Handbook of Philosophy of Imagination*, edited by A. Kind, pp. 355-367. New York: Routledge.

Keen, S. 2007. *Empathy and the Novel*. Oxford: Oxford University Press.

Kimmerer, R. W. 2013. *Braiding Sweetgrass*. London: Penguin.

Lederach, J. P. 2005. *The Moral Imagination: The Art and Soul of Building Peace*. Oxford: Oxford University Press.

Le Guin, U. K. 1988. *The Carrier Bag Theory of Fiction*. London: Ignota.

Lindemann, H. 2001. *Damaged Identities, Narrative Repair*. Ithaca: Cornell University Press.

Linett, M. T. 2020. *Literary Bioethics: Animality, Disability, and the Human*. New York: New York University Press.

MacIntyre, A. 1981. *After Virtue: A Study in Moral Theory*. Notre Dame: University of Notre Dame Press.

McGurl, M. 2012. "The Posthuman Comedy". *Critical Inquiry* 38 (3): 533–553. https://doi.org/10.1086/664550

Morton, T. 2013. *Hyperobjects: Philosophy and Ecology After the End of the World*. Minneapolis: University of Minnesota Press.

Murdoch, I. 2001. *The Sovereignty of Good*. New York: Routledge.

Nussbaum, M. C. 1977. *Cultivating Humanity: A Classical Defense of Reform in Liberal Education*. Cambridge: Harvard University Press.

——— 1990. *Love's Knowledge: Essays on Philosophy and Literature*. Oxford: Oxford University Press.

Ratajczyk, Y. 2023. "Moral Perception as Imaginative Apprehension". *The Journal of Ethics*. https://doi.org/10.1007/s10892-023-09462-5

Smith, A. 2010. *The Theory of Moral Sentiments*. London: Penguin.

Trexler, A. 2015. *Anthropocene Fictions: The Novel in a Time of Climate Change*. Charlottesville: University of Virginia Press.

Vandermeer, J. 2014. *Annihilation*. New York: FSG.

9. Bioethics and (Bio)Art

What is BioArt?

BioArt is an art form dealing with biological material and the relation between human and other-than-human life forms and the environment. BioArtists occupy unique positions, merging the worlds of artistic expression and scientific inquiry and bridging the gap between artist studios and scientific laboratories. They engage with diverse cultural materials, biology, the environment, and other natural sciences. Their work is experimental, co-creating topographies of knowledge and inquiry and addressing controversial topics in science and technology. This provocative approach differs from traditional art's focus on aesthetics.

BioArt's practice consists of working with live tissue, bacteria, living organisms, and life processes to create artworks, where the process itself is often considered art as well. Using scientific processes and practices such as biology, microscopy, or biotechnology (including genetic engineering, tissue culture, and cloning), BioArtists produce their works in laboratories, galleries, or studios and incorporate living, growing, and sometimes unpredictable 'materials' like bacteria, tissues, and even genetically modified organisms.

BioArt extends beyond just working with animals, plants, or microorganisms and encompasses a wide range of artistic practices that engage with the living world. Artists work with animal husbandry, the consequences of climate change, or the notion of empathy towards other-than-human life forms. They might also engage in developing new biomaterials derived from living organisms like fungi, bacteria, algae, and plants to create sculptures, textiles, and other artworks. Another important strand is 'speculative BioArt', which uses speculative designs to imagine future biological technologies and their societal implications through conceptual artworks.

By working with subjects like synthetic biology, tissue culture, and cloning, BioArtists raise questions about the ethical, social, and philosophical implications of these advancements. From its origin in the late twentieth and early twenty-first centuries, BioArt is often intended to highlight themes connected with biological subjects that address or question philosophical notions or trends in science, and can be shocking or humorous at times. Notable bioartists include Eduardo Kac—who created the fluorescent 'GFP bunny', Alba—and Stelarc, who is known for his body-altering projects. Symbiotica,[1] founded by Oron Catts and Ionat Zurr, is a former laboratory for BioArt at the University of Western Australia that helped to institutionalize and ground the work of BioArtists in academia. The scope of BioArt is debated, with some arguing it must involve biological manipulation, while others include work that addresses the social and ethical considerations of the biological sciences. This flexibility in definition pushes the boundaries of what can be considered art.

In sum, by utilizing scientific processes, tools, and even living organisms as mediums, BioArt merges the worlds of artistic expression and scientific inquiry and, therefore, blurs the boundaries between art and science. BioArtistic practice encompasses critical interventions in biotech practices, techno-utopian proposals, and cross-disciplinary exploration. In that sense, BioArt is intrinsically connected with bioethics. These artists grapple with bioethical questions both in their own art and in scientific practice, implicitly addressing our interconnectedness with the more-than-human world.

The ethics of BioArt

The use of living organisms and biological 'materials' as art media raises questions about the moral status of these entities and what kind of ethical considerations should be applied to them. This is generally attributed to a post-humanist view where the argument that 'species is not important' prevails. Post-humanism challenges the anthropocentric view that humans are superior to and separate from other species. Instead, it promotes a more egalitarian and interconnected perspective on relationships between humans and other-than-humans.

Following this ethos, BioArtists take responsibility for various life forms: humans, other-than-human animals, organs, cells, and bacteria. They challenge traditional humanist ethics by recreating, pushing, and remoulding life. The protection of life's unfolding is central to this ethics, along with critical reflection on emergent life forms. Some ethical questions may be unprecedented, deepening our understanding of ethical issues in BioArt. Discussions about these ethical issues often fall within the framework of bioethics. An integrated approach here inspires novel ways of thinking about ethics in art and technology. If we explore some implications of using live organisms as art material, we can distinguish between:

1. BioArt and modification: BioArt involves working with bacteria, live tissues, or other organisms by modifying life processes to create art. The ethical dilemma here is that artists must consider the impact of altering living beings for artistic expression. Is it ethical to manipulate life in this way?

2. Dominion over life: Encoding human dominion over life challenges our role as creators and stewards.

---

1 https://www.symbiotica.uwa.edu.au

3. Nature's pushback: Dominion does not always persist despite human intervention. Bacteria can exchange genetic information, resisting our control. How do we balance artistic expression with respect for the autonomy of living organisms?

This use of living organisms in BioArt can be misinterpreted or sensationalized, leading to censorship and controversy. However, artists argue that they are simply making visible what is already happening as biological research behind closed doors in laboratories. For example, disposing of living tissue at the end of BioArt exhibitions is an ethical challenge. Artists often have a 'killing ritual' to expose the tissue to the environment and contaminate it, engaging the audience in ethical decision-making.

BioArt and bioethics

Beyond ethical questions raised by the practice of BioArt itself, BioArtists can and often do inspire more thorough engagement with conceptual and ethical questions in contemporary sciences. Integrating longstanding frameworks in bioethics with existing discourses on art and morality can create new approaches tailored to the subject matter. Several aspects of BioArt's practice directly inspire novel ways of engaging with longstanding issues in the (life) sciences and bridging the gap between armchair ethics and society.

Involvement of the audience

BioArt aims to engage the audience in a multidimensional way, making the audience participate in the artwork itself. By combining striking sonic, visual, olfactory, tactile, and thought-provoking works and concepts, BioArtists seek to stimulate both the intellect and the senses. Debates around issues such as the use of animals, genetic modification, and the creation of new life forms are leading directly to discussions of bioethics. The affective and visceral qualities of BioArt may spur audiences to adjust, revise, or develop their personal ethical frameworks. In this way, BioArt can provide novel impulses for the evolution of bioethical thinking and practices and may provide feedback on or influence bioethical frameworks.

BioArt projects can explore and assess the societal and ethical impacts of emerging biotechnologies. By engaging the public and provoking discussion, BioArt can function as a form of 'material technology assessment' and reflection on the implications of possible future developments.

BioArt as artistic research

Many BioArtist practices are situated in the broader field of artistic research—a swiftly evolving niche in the arts that de Assis (2020) argues that Artistic Research merges the contrasting activities of the artist and the researcher. Whereas the artist is concerned with making, imagination, experience, sensation, and the 'subjective production of new relationships', the researcher focus is on analyzing, measuring and giving meaning towards objectively articulating new knowledge. Artistic Research, therefore, lies at the intersection between making and analyzing, bringing together two distinct modes of operation. BioArt particularly blurs the boundaries between art, science, and ethics in epistemically fruitful ways, prompting us to reflect on our responsibilities as creators and inspiring us to take up novel perspectives.

Bioethics, Biology, BioArt

More common, fertile ground can be found in each field's entanglement with biology. For bioethics, this intimate connection with all life should permeate all levels of scientific practice from the design phase of research. Kristien Hens (2022) describes bioethics as aproper ethics of life, a discipline in times of super wicked problems that includes thinking about the lives and health of humans and other-than-human beings, the macrocosm and the entanglements of all these entities. BioArtists often collaborate with scientists and ethicists to navigate the ethical complexities of their work. This interdisciplinary approach can lead to new understandings and approaches at the intersection of art, science, and ethics, providing methods to make these issues tangible. BioArtists, like bioethicists, use concepts of life to guide their practices, and many share the same view of life. Many work to some extent in an institutional scientific setting, and many share critiques on science that align to some extent with the main points of the slow science movement (see, for example, Isabelle Stengers below). Both bioethicists and BioArtists inquire about the environment, new technologies, and health in the broad sense of the term, and both might become embedded in a scientific research setting. Bioethics, science, and BioArt are part of a dynamic dialogue that navigates the boundaries of life, creativity, and responsibility. Let us now turn to some methodologies from the art world that can help us think in terms of bioethics.

How BioArt can contribute to ethical considerations towards the 'unknown'

BioArt can rephrase questions to help us reconsider what is outside of our moral view, and can also suggest scientific methodologies that do not rely on dualisms, objectification, species hierarchies, and extractivism.

The methodologies used in BioArt might give tools that can help us further. BioArt attempts to acknowledge 'otherness' on its own terms. In practice, this often implies ways to deeply sense and (cor)respond—or *attune* and explore principles of attuning as a possible tool for furthering critical thinking on post-anthropocentric futures and 'staying with the trouble'. Attunement cannot make the unknown completely knowable, but it may offer opportunities to establish relationships with what is unknown and appreciate the unknown for what it is.

Dominant ethical theories and principles are not always well equipped for working with abstract or unknown parts of our environment. An example is the ethics of the alien deep sea that we can only access indirectly. This invisibility and alienness of the sea and its inhabitants can provide new insights into ethical methodologies. Rather than tightening our grip on what is already directly known, considering what is outside our usual moral view may be just as valuable. This means reflecting on how to make ethical judgements in light of what is still speculative. To help us do so, art may stimulate interest in these unknown and invisible parts of the ecosystem and, thus, pave the way towards including the unknown in ethical reflection.

Time

As Isabelle Stengers points out in her *Slow Science Manifesto* (2018): it is a matter of unlearning an attitude of more or less cynical ('realist') resignation and becoming sensitive once again to what we perhaps know, but only as in a dream. It is here that the word 'slow' is apt. Speed demands and creates an insensitivity to everything that might slow things down: the frictions, the rubbing, the hesitations that make us feel we are not alone in the world. Slowing down means becoming capable of learning again, becoming acquainted with things again, and reweaving the bounds of interdependency. It means thinking and imagining and, in the process, creating relationships with others that do not try to capture them but rather leave them be. It means, therefore, creating among us and with others the kind of relation that defines a life worth living and the knowledge worth being cultivated.

Ephemeral art

BioArt is often ephemeral, embracing impermanence and creating fleeting, fragile pieces that might invite contemplation about our impermanent existence. Artists intentionally use materials with limited durability—such as ice, light, leaves, water, steam, electricity, and radiation—which change over time and challenge traditional notions of permanence. Temporary installations and performance (Bio)Art are unique to the time and place of their existence, emphasizing the transient essence of life itself. Audiences are invited to witness the ongoing processes of growth, transformation, and decay, as well as the themes of nature, mortality, and the passage of time. As such, BioArt can yield new insights for debates about the ethics of environmental restoration or end-of-life care.

Documenting

Ephemeral art, marked by its temporary and transitory nature, presents a captivating challenge for artists seeking to preserve and document their work. Archival methods like photography and video, for example, can record the actual event. This is a subjective option to reflect the artist's viewpoint. Alternative archival methods might include written accounts, publications, posters, and even manifestos associated with the artwork. Here, both the event and its documentation are subject to the vagaries of time and memory. This can inspire new ways of thinking in bioethics, particularly in contexts like medical storytelling, lived experience, and end-of-life care, where ephemeral and subjective narratives play a crucial role in shaping ethical understanding and practice.

Conclusion

We examined the intersection of BioArt and bioethics, showing how artistic practices that engage with living materials and biotechnologies can open up new ethical perspectives. BioArt challenges traditional boundaries between science and art by working with organisms, tissues, and biological processes, often raising provocative questions about the moral status of life forms and our responsibilities toward them. Rooted in post-humanist thought, it questions human exceptionalism and fosters reflection on our entanglements with other-than-human life. By involving audiences emotionally and intellectually, BioArt can influence personal ethical frameworks and contribute to public dialogue on emerging biotechnologies.

Bibliography

Berger, Erich, Kasperi Mäki-Reinikka, Kira O'Reilly, and Helena Sederholm, eds. 2020. *Art as We Don't Know It*. Espoo: Aalto ARTS Books.

Braidotti, Rosi, and Maria Hlavajova, eds. 2018. *Posthuman Glossary*. London: Bloomsbury.

de Assis, Paulo, and Lucia D'Errico, eds. 2019. *Artistic Research: Charting a Field in Expansion*. Lanham: Rowman & Littlefield.

Hannula, Mika, Juha Suoranta, and Tere Vadén. 2005. *Artistic Research. Theories, Methods and Practices*. Helsinki: Academy of Fine Arts.

Hens, K. 2022. *Chance Encounters. Exploring the Ethics of Life*. Cambridge: Open Book Publishers, https://www.openbookpublishers.com/books/10.11647/obp.0320

Ingold, T. 2021. *Correspondences*. Cambridge: Polity.

Lipari, L. 2014. *Listening, Thinking, Being: Toward an Ethics of Attunement*. Pennsylvania: Penn State University Press.

Stengers, I. 2018. *Another Science is Possible: A Manifesto for Slow Science*. Melford: Polity.

Index

act utilitarianism 17
Adam, Alison 58
ageism 62
Agent Orange 2
algorithmic bias 53, 56
androcentrism 32
animal experiments 73, 77
animals 3, 5, 7, 17, 20, 23–24, 31, 36, 73–77, 84–85, 100, 132–133, 135, 141
Anscombe, Elizabeth 21
anthropocentrism 29, 31–32, 128, 142
anthropology 12
antibiotic resistance 35
antibiotics 35–37
applied ethics 1, 7, 11, 15, 128
Aristotle 21–22, 36, 74
artificial intelligence (AI) 20, 53–58
   generative AI 55
   trustworthy AI 53–54
autonomy 11, 18, 19, 24, 35, 48, 49, 50, 51, 52, 53, 54, 55, 57, 59, 60, 64, 104, 114, 119, 141. *See also* parental autonomy
Aztec 22

Baindur, Meera 121
Barad, Karen 4
Barnes, Elisabeth 61
Beauchamp and Childress 48–50, 58
beneficence 11, 48, 50, 51, 59, 114. *See also* non-maleficence; *See also* procreative beneficence
Bentham, Jeremy 16–17, 74
bikini medicine 67
BioArt 140–143
Blake, William 130
Brahman 121
Brethren of Purity 132
Buen Vivir 11, 39–40
business ethics 7

Callicott, John Baird 31
care ethics 7, 11, 15, 22–24, 48–50, 61
categorical imperative 18–20

cell factories 110–111, 117
cell therapy 109
Charon, Rita 49
climate change 17, 34, 36–38, 130, 133
clinical decision-support systems 57
clinical equipoise 66
clinical ethics 50
Coetzee, J. M. 132
collective responsibility 102
consequentialism 16
Costello, Elizabeth 132
COVID-19 61, 63
CRISPR/CAS9 59
critical race theory 19
Cupples, Laura 63
Cutas, Daniela 100, 102

Darwin, Charles 74
data portability 69
decolonizing ethics 11
deep ecology 32
deontology 15, 18–19, 21, 75, 113
Descartes, René 74
descriptive ethics 6
de Waal, Frans 7
Dewey, John 127
D'Ignazio, Catherine 57
dignity 18, 50, 54, 57
disability 20, 60–61, 63, 131
distributive justice 36–37
DNA methylation 95
duty 1, 7, 18, 20, 21, 48, 73, 74, 75, 99, 121. *See also* beneficence

ecofeminism 24, 32
ecofiction 130
egalitarianism 32, 101
egoistic prudence 7
embedded ethics 11
embryo selection 58–59
empirical bioethics 11
empowerment 104

enlightenment 16
environmental ethics 1, 31, 38, 75
environmental fascism 31
epigenetics 1, 59, 94, 95, 97, 99, 100, 101, 102, 103. *See also* DNA methylation
epistemology 4–5
ethical absolutism 15
Ethica Nicomachea 21
ethico-onto-epistemology 4
ethics dumping 115
eudaimonia 21–22
eugenics 60, 119
experience machine 16
experimental philosophy 10
explainability 54, 57
explicability 53–55
extractivism 39

fairness 7, 37, 53–56, 118–119
fair subject selection 66
favourable risk-benefit ratio 66
feminism 4, 5, 6, 11, 22, 24, 32, 48, 52, 58, 60, 63, 101, 129. *See also* ecofeminism
feminist philosophy of science 11
fiction 127, 128, 129, 133. *See also* ecofiction
field philosophy 12
flourishing 24
Foot, Philippa 16
Frank, Arthur 49
Fricker, Miranda 57
future generations 36–37, 102, 119

Galston, Arthur 2
gender 24, 48, 54, 56–57
gene editing 58, 109
General Data Protection Regulation (GDPR) 3, 69
Gettier cases 9–10
Gettier, Edmund 9
Ghosh, Amitav 133
Gilligan, Carol 7, 22

Haraway, Donna 129
Havasupai Tribe 68–69
healthcare ethics 36
hedonism 16
hedonistic calculus 16
helicopter research 115
Hinduism 120, 122
Hobbes, Thomas 7

Human Genome Project 6
hybrid approach 84–85
hyperobject 133

implicit bias 56
independent review 66–67
indigenous knowledge 24, 129
inductive risk 6
informed consent 49–52, 57, 66–69
integrity 31, 49–50, 54
interdependency 40, 143
intuition 8–11, 60–61, 67
Ishiguro, Kazuo 127

Johnson, Gabbrielle 57
justice 7, 11, 21, 34, 35, 36, 37, 38, 50, 53, 55, 56, 60, 63, 67, 95, 114, 123, 132. *See also* distributive justice; *See also* fairness
justified true belief 9

Kant, Immanuel 18–21, 75
Katz, Eric 31
Keen, Suzanne 128
Kichwa 39
Kimmerer, Robin Wall 129
Kittay, Eva 24, 60
Klein, Lauren 57
Kohlberg, Lawrence 7, 22
Korsgaard, Christine 19
Kossakovsky, Victor 132
Kuhn, Thomas 5

Lacks, Henrietta 68
Le Guin, Ursula 129–130
Leopold, Aldo 31
Lindemann, Hilde 132
Linett, Maren 128, 131

MacIntyre, Alasdair 21
media ethics 7
medical paternalism 50, 58
Meijer, Eva 131
metabolic engineering 111
metaethics 7
metaphors 111–112
Midgley, Mary 76
Mill, John Stuart 16
Mills, Charles 19
minimal genome 110
monism 15. *See also* non-dualism
Moore, G. E. 8

moral law 20
moral theories 15, 61, 74, 76
Morton, Timothy 133
Murdoch, Iris 127

Naess, Arne 32
narrative ethics 49–50, 132
naturalism 8
nature 7, 32, 39–40, 49–50, 95, 102, 110, 112–114, 120–123, 127, 130, 143
nested communities 31
neurodiversity 4
Noddings, Nell 23
non-dualism 120, 122. *See also* monism
non-maleficence 11, 48, 50, 51, 53, 55, 114. *See also* beneficence
normative ethics 7
Nozick, Robert 16
Nussbaum, Martha 127

One Health 35
O'Neill, Onora 3, 51
ontology 4
orthogonal biosystems 110

paradigm 5
parental autonomy 104
Parfit, Derek 9
paternalism 50, 64. *See also* medical paternalism
patient-physician relationship 50
personal identity 9
phronesis 21
Plato 36
political philosophy 4
postcolonialism 48
posthumanism 24
Potawatomi 129
Potter, Van Rensselaer 1
practical philosophy 4
precautionary principle 67
Preimplantation Genetic Testing 59
prima facie 48
principlism 48–50
prioritarianism 63
procreative beneficence 59
procreative liberty 59
public health 35, 53, 61–64, 98–99
Puig de la Bellacasa, Maria 24

quality-adjusted life years (QALYs) 63

race 19, 48, 55, 56, 57, 76. *See also* critical race theory
Railton, Peter 8
reciprocity 40
Regan, Tom 20, 31, 75
regulatory circuits 111
relationality 24
relativism 48
remediation 37, 109
replaceability 29–30
reproductive ethics 58, 59. *See also* procreative beneficence; *See also* procreative liberty
research ethics 64. *See also* fair subject selection
respect for enrolled subjects 66
responsibility. *See also* collective responsibility
restitution 30
rights 18, 20, 23, 40, 53, 55, 60, 73–75, 85, 104, 114, 127, 132
Ritalin 9
Robertson, John 59
rule utilitarianism 17
Ryder, Richard 76

Sanatana Dharma 121–123
Savulescu, Julian 59
scientific knowledge 5
scientific validity 66
Scully, Jackie Leach 60
semi-synthetic artemisinin 118
sentiency 29
separation 1, 74, 111, 122
Singer, Peter 17–18, 75–76
slippery slope 112–113, 120
social contract 7
social philosophy 4
sociobiology 31
solarpunk 130
speciesism 17, 76. *See also* anthropocentrism
Stengers, Isabelle 142–143
substituted judgement 51
sumak kawsay 39
supererogatory 18
Swift, Taylor 20–21
synthetic biology (SynBio) 109–114, 120, 123

telos 22
Tengland, Per-Anders 64
therapeutic orphans 67

thin moral theory 22
Thoreau, Henry David 130
thought experiments 9–10, 132
tiered consent 69
transparency 54–55, 57
triage 62–63
trolley problem 16
Tronto, Joan 23–24
Tuskegee syphilis experiment 65

universal law 19–21
utilitarianism 15, 16, 17, 18, 19, 21, 74, 75. *See also* act utilitarianism; *See also* rule utilitarianism

validity 84

van Aquino, Thomas 21
Vandermeer, Jeff 130
virtue 15, 20–24, 36, 128
virtue ethics 15, 20–24
Vishnu 123
vulnerability 23–24, 50, 66, 130

Wilson, Shawn 122
Wordsworth, William 130
Wu Ming-Yi 131

Yasuni National Park 38, 39, 40

About the Team

Alessandra Tosi was the managing editor for this book.

Annie Hine proof-read the manuscript and compiled the index.

Jeevanjot Kaur Nagpal designed the cover. The cover was produced in InDesign using the Fontin font.

Cameron Craig typeset the book in InDesign. The main text font is Tex Gyre Pagella and the heading font is Californian FB.

The conversion to HTML was performed with with epublius, an open-source software which is freely available on our GitHub page at https://github.com/OpenBookPublishers.

Jeremy Bowman created the EPUB and PDF editions.

Laura Rodríguez Pupo was in charge of marketing.

This book was peer-reviewed by Henk ten Have, Professor Emeritus at the Center for Healthcare Ethics at Duquesne University in Pittsburgh, USA.

Experts in their field, our readers give their time freely to help ensure the academic rigour of our books. We are grateful for their generous and invaluable contribution

This book need not end here...

Share

All our books — including the one you have just read — are free to access online so that students, researchers and members of the public who can't afford a printed edition will have access to the same ideas. This title will be accessed online by hundreds of readers each month across the globe:
why not share the link so that someone you know is one of them?

This book and additional content is available at
https://doi.org/10.11647/OBP.0449

Donate

Open Book Publishers is an award-winning, scholar-led, not-for-profit press making knowledge freely available one book at a time. We don't charge authors to publish with us: instead, our work is supported by our library members and by donations from people who believe that research
shouldn't be locked behind paywalls.

Join the effort to free knowledge by supporting us at
https://www.openbookpublishers.com/support-us

We invite you to connect with us on our socials!

BLUESKY
@openbookpublish
.bsky.social

MASTODON
@OpenBookPublish
@hcommons.social

LINKEDIN
open-book-publishers

Read more at the Open Book Publishers Blog
https://blogs.openbookpublishers.com

You may also be interested in:

Towards an Ethics of Autism
A Philosophical Exploration
Kristien Hens

https://doi.org/10.11647/obp.0261

Ethics for A-Level
Mark Dimmock and Andrew Fisher

https://doi.org/10.11647/obp.0125

Beyond Price
Essays on Birth and Death
J. David Velleman

https://doi.org/10.11647/obp.0061

www.ingramcontent.com/pod-product-compliance
Lightning Source LLC
Chambersburg PA
CBHW041241240426
43668CB00023B/2449